Text Book of
Agricultural Heritage

Authors

Dr. Utpal Giri
Dr. Md. Hedayetullah

SCIENTIFIC PUBLISHERS

Published by:

SCIENTIFIC PUBLISHERS (INDIA)
5A, New Pali Road, P.O. Box 91
Jodhpur 342 001, India
E-mail: info@scientificpub.com
Website: www.scientificpub.com

Print 2024

ISBN : 978-93-89832-82-2
eISBN: 978-93-89412-53-6

Printed in India

FOREWORD

Agricultural heritage refers inherited values carried from one generation to other generation. Agricultural heritage indicates the values and traditional practices and methods adopted in ancient India, which are more relevant for present day system for maintaining sustainable agriculture. The objectives of the book are to increase awareness and implant a sense of pride amongst the people of the rich unique heritage of Indian agriculture, to know as our agriculture has sustainable practices for generations and to stimulate scientific research based on traditional technology.

The present book "Text Book of Agricultural Heritage" covers thirteen chapters, providing in depth description of key contents, issues and topics. The first chapter introduces the agricultural heritage, the second chapter focuses on ancient agricultural practices, third chapter explains the past and present status of agriculture and farmers in society, fourth elaborates history of agricultural development, fifth describes indigenous technology knowledge in agriculture (ITKs) and allied agricultural sector, sixth is devoted to crop voyage. The scope of agriculture scope, importance of agriculture and agricultural resources available in India have been explained in the seventh chapter, eighth is devoted to crop significance and classifications, ninth chapter highlights the national agriculture setup in India, tenth describes the current scenario of Indian agriculture, eleventh chapter is related to Indian agricultural concerns and future prospects. Chapter twelve focuses on Indian Council of Agriculture Research (ICAR) institutes and chapter thirteen focuses on different types of model questions.

The books has got true reflections of 5th Dean's Committee recommendations in giving the syllabuses for the students in under graduate level to the State Agricultural Universities in the country and legacy of referral works in continuation and to follow up present day modern agriculture.

The book would come to help with much contribution to the students, researchers, planners and policy makers throughout the country and abroad.

Professor A Zaman
Visiting Professor & Advisor Agriculture,
The Neotia University, Kolkata Campus at Sarisa-743368,
24 Paraganas (S), West Bengal
Co-founder, Vice President, and Member Secretary of
International Institute of Bengal Basin, H.Q based in Berkeley, California, USA
Formerly Consultant to World Bank, Minor Irrigation Project
Head, Irrigation Expatriate Team, Government of Uganda, East Africa
Head, Department of Agronomy, BCKV (2008-2012)
Dean, Faculty of Agriculture, Chief Scientist, Water Management, Director of Research
Bidhan Chandra Krishi Viswavidhayalaya (BCKV) West Bengal India.

PREFACE

Agriculture is our backbone of the society, way of life and not an occupation. History is defined as the continuous record of past events, where as heritage denotes the inherited values carried from one generation to other generation. Agricultural heritage indicates the values and traditional practices and methods adopted in ancient India, which are more relevant for present day system for sustainable agriculture. The objectives of the book are to increase awareness and implant a sense of pride amongst the people of the rich unique heritage of Indian agriculture, to know as our agriculture has sustainable practices for generations and to stimulate scientific research based on traditional technology.

This book is intended as a text book for under-graduate students in agriculture. It will be a valuable reference to professional faculty members, agricultural scientists, and students.

Citations are avoided in the text so that the reader is not distracted. At the end of the book, a list of references has been given from which most of the material in the book is drawn. Theses references will also help students for further reading.

The author acknowledge their indebtedness to the authors of various books, bulletins, monographs, periodicals and published papers from which most of the material in this book is drawn. The authors are indebted to several individuals and organizations for the supply of useful material.

The book is spread over 13 chapters, providing in-depth description of key contents, issues and topics. The first chapter introduces the Agricultural Heritage, the second chapter focuses on ancient agricultural practices, third chapter explains Past and present status of agriculture and farmers in society, fourth elaborates history of agricultural development, fifth describes indigenous technology knowledge in agriculture and allied agricultural sector, sixth is devoted to Crop Voyage. Agriculture scope, Importance of agriculture and agricultural resources available in India have been explained in seventh chapter, eighth is devoted to Crop significance and Classifications, ninth chapter highlights the National agriculture setup in India, Tenth describes the current scenario of Indian agriculture, eleventh chapter is related to Indian agricultural concerns and future prospects.

Utpal Giri

Md. Hedayetullah

Contents

CHAPTER 1

Agriculture Heritage in India

Agriculture is our backbone of the society. History is defined as the continuous record of past events, where as heritage denotes the inherited values carried from one generation to other generation. Agricultural heritage indicates the values and traditional practices and methods adopted in ancient India, which are more relevant for present day system for sustainable agriculture.

Globally Important Agricultural Heritage Systems (GIAHS), as defined by the FAO (Food and Agriculture Organization of the UNO), are: "Remarkable land use systems and landscapes which are rich in globally significant biological diversity evolving from the co-adaptation of a community with its environment and its needs and aspirations for sustainable development". Worldwide, farmers developed, modified and maintained specific agricultural systems, landscapes and herders based on available diversified natural resources, using locally adapted management practices. These local knowledge and experiences on ingenious agricultural systems reflects the humankind evaluation, knowledge diversity and its profound relationship with nature.These systems have resulted not only in outstanding landscapes, maintenance and adaptation of agriculturally significant biodiversity, ITK systems and ecosystems, but, above all, supply of sustainable multiple goods and services, food and livelihood security for thousands of local people, well beyond their borders and conserved the natural resources and biodiversity.

Agricultural heritage systems can also be found throughout the world with an area of 5 million hectares. These vast heritage serves as a important combination of social, cultural, ecological and economical services to humankind. The existence of various GIAHS through out the world testifies to the inventiveness and ingenuity of people in their use and management of the these natural resources, biodiversity and ecosystem dynamics, and ingenious use of physical attributes of the landscape, codified in traditional but evolving knowledge, practices and technologies. Whether it has been recognized by the scientific community, these traditional agricultural systems build up the foundation for contemporary and future agricultural innovations and technologies. Their cultural, ecological and agricultural diversity is still found in many parts of the world, and considered as unique methods of agriculture. Through a remarkable process of co-evolution of Humankind and Nature, GIAHS have represented the accumulated experiences of local peoples which have emerged over centuries of cultural and biological interactions and synergies.

Agriculture Heritage in India

Indian heritage is unique and prestigious than any other civilization in this world. We must feel proud being a citizen of India for its rich cultural heritage. Agriculture in India has a long history from Neolithic age of 7500-4000 B.C. People changed their life style from nomadic hunter of wild berries to cultivator of land. Agriculture is gifted from the wisdom and teachings of great saints. The knowledge achieved and techniques adopted have been passed down through generations. The traditional farmers have developed the eco-friendly farming systems and mixed farming, mixed cropping, crop rotation etc type of systems were practiced by them. The older generations of the farmers of India passed the depth of knowledge and experience of great epics of ancient India.

The most ancient literature available in India was "Vedas" which includes the four Vedas (rig, yajur, sama, atharvana), nineteen Brahmanas (two associated with the Rigveda, six with the Yajurveda, ten with the Samaveda and one with the Atharvaveda). Some other important ancient epic literatures are Aranyakas, Sutra literature, Susruta Samhita, Charaka Samhita, Upanishads, the epices Ramayana and Mahabharata, Puranas (20), Buddhist and Jain literature, and texts such as Krishi-Parashara, Kautilya's Arthasastra, Panini's Ashtadhyayi, Sangam literature of Tamils, Manusmirti (laws), Varahamihira's Brihat Samhita (maths & astrology), Amarkosha, Kashyapiya-Krishisukti and Surapala's Vriskshayurveda. These literatures were most likely to have been composed between 6000 BC to 1000 AD. The information related to the biodiversity and agriculture (including animal husbandry) are available in these texts.

Rigveda, which is the most ancient literary work of India, believed that Gods were the foremost among agriculturists.According to the thesaurus, "Amarakosha" which was written by the Jain or Buddhist scholar Amarasimha in Sanskrit, Aryans were agriculturists. Agriculture, cattle rearing and commerce were the essential subjects of the king that must be learnt. This was prescribed by Manu and Kautilya.The great compiler of the Yoga Sūtras, "Patanjali" mentioned that the economy of the country depended on agriculture and cattle-breeding. According to 'Puranas', the ancient Indians had intimate knowledge on all agricultural operations.The books dealing exclusively with techniques of agriculture were Sage Parashara's 'Krishiparashara' in 1000 A.D. Other important texts are Agni Purana and Krishi Sukti written by Kashyap (500 A.D.).Some of the well known ancient classics of India are namely, Kautilya's'Arthashastra'; Patanjali's 'Mahabhasya'; Panini's 'Astadhyayi'; Amarsimha's 'Amarkosha'; Varahamihira's 'Brahat Samhita'and Encyclopaedic works of Manasollasa. These classics literary works indicated the knowledge and wisdom of the people of ancient period.Ancient works of Tamil and Kannada contain lot of useful information on agriculture in ancient India. Indian agriculture made tremendous progress in the rearing of cows and buffaloes, sheep and goats, cultivation of trees and shrubs, spices and condiments, food and non-food crops, fruits and vegetables and also developed nature friendly farming practices. These practices had done with the social and religious relationship and it was the way of life for the people. Domestic rites and rituals of festivals often coincided with the four main agricultural operations like ploughing, sowing, reaping and harvesting.

In the Rigveda, it has been mentioned about the hundreds and thousands of cows, horse drawn chariots and race courses where chariot races were held; camels yoked to the chariots; sheep and goats offered as victims for worship, and the use of cloths made from wool.The famous Cow Sukta indicated that the cow also became the basis of rural economy. Sukta is also worshiped as the mother of the Vasus, the Rudras and the Adityas, and also the pivot of Immortality. The Vedic Aryans occupied large forests at their disposal for collection of timber. The physicians of that age reared plants and herbs for medicinal purposes, as appears in the Atharva Veda. Parjanya was worshiped as the god of rain. Agriculture was solely depended upon the favours of Parjanya. His thunders are described as food-bringing. The farmers' vocation was held in high regard.

More than 75 plant species was mentioned in the four Vedas, Satapatha Bhrahamna mentioned over 25 species, and more than 320 plants were found in Charkaa Samhita (300 BC) an Aayurvedic (Indian medicine) treatise. Over 750 medicinal plant species were recorded in Susruta (400 BC). Rigveda (4000 BC), the oldest book, mentions a large number of poisonous and non poisonous aquatic and terrestrial, and domestic and wild creatures and animals. About 500 species of plants were written in Puranas. The science of arbori-horticulture had developed well and has been documented in Surapala's Vrikshayurveda. From the age of Vedas, forests were given more importance and protected for ecological balance. Kautilya in his Artha Sastra (321-296 BC) provides a long list of trees, varieties, of bamboos, creepers, fibrous plants, drugs and poisons, skins of various animals, etc., that come under the purview of forest officer. He also mentioned that superintendent of forests had to collect forest produce through the forest guards.

According to Manu (Manusmriti, 2nd Century BC), the preservation of wild animals was maintained and hunting as a sport by the ruler was considered as detrimental to proper development of the character and personality. In the Puranas (300-750 AD), the names of the experts in animal husbandry were found like Shalihotra on horses and Palakapya on elephants. For instance, treatment of animal disorders was mentioned in Garudapurana while Aswashastra is dealing with the classical work on the treatment of horses. One chapter in Agnipurana is written about the treatment of livestock and another on treatment of trees.

Need and Importance for Studying Agricultural Heritage

Our agriculture has lot of inherited sustainable practices those are orally passed from one generation to other generation. Agriculture in India is not only an occupation; it is the backbone of the society and the way of life for many people of India. Hence it is important to the present day generation to become aware about our ancient and traditional agricultural systems and practices. This will help us to develop the future scope of research strategy also.

India has made tremendous improvement in agriculture and its allied sectors, but due to the intensive use of inputs without considering their adverse effects to the environment on a long term basis which has created several problems related to sustainability of agriculture. Non-judicious application of chemical fertilizers, insecticides and irrational use of natural resources is threatening the agro eco systems. Soil is degraded, water and air are polluted and plant and animal genetic resources are in danger. Therefore, importance is now shifted to sustainable form of agriculture.

The indigenous technical knowledge (ITK) is the valuable information for the sustainable agriculture, because these technologies have been carried on from one generation to another as a family technology. There are several examples of valuable traditional technologies in India but unfortunately these small local systems are going to extinct day by day. It is imperative that the agricultural scientist should collect, document and analyze these technologies so that the scientific principle/basis behind them could be properly understood. Once this done, further refinement and upgradation by blending them with the modern scientific technology will be easier for us.

Objective of the Course

- Agriculture in India - Backbone of the society, Way of life and not an occupation
- To increase awareness and implant a sense of pride amongst the people of the rich unique heritage of Indian agriculture
- To know as our agriculture has sustainable practices for generations.
- To stimulate scientific research based on traditional technology.

India'a Agricultural Heritage Status in Press World Heritage Status for India's Farming Techniques—The Hindu, FEB 13, 2010

India's locally developed farming techniques look set to take their place on the world heritage map alongside the country's national parks of outstanding beauty and its grand monuments to culture. Unlike Konarak, Kaziranga and the Taj Mahal, the rice crop of Koraput, the salt water farms of Kuttanad, and the paddy fields of Thanjavur could receive the same importance due to unique agricultural systems in an era of climate change under an initiative of the UN Food and Agriculture Organisation (FAO). "These sites are protecting our food security. They are our heritage... The techniques were developed by farmers, not by scientists or anything else. The technology is their own," said M.S. Swaminathan, Chairman MSSRF, to the reporters prior to an international conference on biodiversity at the Research Foundation.

Under the changing climatic conditions, Globally Important Agricultural Heritage Systems (GIAHS) are natural solutions in the future and the natural evolution of farming. The Koraput region of Orissa in India has been nominated for the first candidate of GIAHS status.The tribal groups of that region has developed the variety of rice, millets, pulses, and medicinal plants using traditional cultivation practices. Dr. Ajay Parida, Executive Director, MSSRF mentioned about the farmers of Kuttanad region in Allapuzha, Kerala, for their production of crops in sea water and the 2,000-year-old system of irrigating paddy in Thanjavur region. These papers have been submitted to the FAO.

So far, five systems from five different countries have been identified as GIAHS. These are - Andean agriculture in Peru, Chiloe agriculture in Chile, the Ifuago rice terraces of the

Philippines, the Magreb Oases in Algeria and the Upland pastures that cross the borders of Kenya and Tanzania.

Kudos for Koraputs Agricultural Heritage—Indian Express Jan 08, 2012

United Nation's Food and Agriculture Organisation (FAO) has identified the efforts of the tribal community of Koraput to conserve biodiversity and develop climate resilient agricultural activities as a Globally Important Agricultural Heritage System (GIAHS).

The recognition of the Koraput traditional agricultural system as a GIAHS site is expected to guarantee local and international efforts for the conservation of biodiversity and sustainable use of its genetic resources.

GIAHS defined land use systems and landscapes are rich in globally significant biological diversity which are developed from the co-adaptation of a community with its environment and its needs and aspirations for sustainable development.

Mentioning this at the inauguration of the 99th Indian Science Congress here, Prime Minister Manmohan Singh lauded the Koraput tribals for the achievement.

The Koraput region is rich in globally important agricultural biodiversity. Theses agro-biodiversity includes, 340 landraces of paddy, eight species of minor millets, nine species of pulses, five species of oilseeds, three species of fibrous plants and seven species of vegetables.

Another region which is rich in genetic resources of medicinal plants is the Jeypore region. It has been observed of about more than 1,200 medicinal plant species in this region. Some of the endemic plant species of the region are used for curing different diseases including gastrointestinal disorders, malaria and bone fracture which is the concern to our ecosystem.

The tribal groups also contributed to biodiversity and knowledge systems through their rich traditional knowledge about forest species.

Asian has been considered as the centre of origin of aus ecotype of rice because of its genetic diversity. The landraces or traditional varieties growing in this region are considered to have dominant genes for biotic and abiotic stresses, aroma and palatability, and it has been thought that these are the genetic resources in future plant breeding and biotechnology programmes to mitigate the adverse effect of climatic and environmental changes. With the help of their traditional knowledge, the tribal and rural families of this area have been developing and conserving these genetic resources by careful and continuous selection under the changing environment and agricultural practices, but these merits have not been recognised which they deserve.

Indian Agriculture

Indian agriculture began by 9000 BCE as a result of early cultivation of plants, and domestication of crops and animals. Later improved agriculture had been developed with implements and techniques. Double monsoons led to double cropping sequence in one year. Indian products soon reached to the other parts of the world with the help of existing trading networks and also foreign crops were introduced to India. Plants and animals were considered essential to their survival by the Indians and came to be worshiped and venerated.

In the middle ages, sophisticated agriculture was done with the development of irrigation system in India and Indian crops affecting the economies of other regions of the world under Islamic patronage. Land and water management systems were developed with an aim of providing uniform growth. Despite some stagnation during the later modern era the independent Republic of India was able to develop a comprehensive agricultural program.

CHAPTER 2

Ancient Agricultural Practices, Relevance of Heritage to Present Day Agriculture

Kautilya's Artha-Sastra, Agriculture, Animal Husbandry, Commodity Trade etc. – Features of Village

Kautilya (popularly known as Vishnugupta or Chanakya) (321-296 BC) was a great scholar of time. He wrote most famous a treatise titled, Artha-sastra, which deals with the management of resources. During that time, agriculture, breeding of cattle and trade were grouped into a science called Varta. Kautilya gave great importance to agriculture and suggested a separate post of head of agriculture designated as Sitadhakashya. Agriculture now-a-daysis receiving prime importance, by policy and administrative support from government like i) Supply of good seeds and other inputs ii) Provision of irrigation water iii) Forecasting of rainfall by IMD iv) Financial support to purchase of machineries v) Marketing and safe storage. Kautilya mentioned all these important aspects in his book. He advised many important aspects in agriculture which are highly relevant now – a days. Some of these mentioned aspects are discussed below:

1. The superintendent of agriculture should be a knowledgeable person in agriculture and horticulture. There was a provision to appoint a person who was not an expert but he was assisted by other knowledgeable person. These type of appointments are still follow even in today, like appointment of the directors of agriculture, horticulture and sometimes civil servants are selected as technical persons.
2. Hiring of labours by land owners before sowing: Slaves and prisoners were organized to sow the seeds in time. He also emphasized to thorough ploughing that provides good soil texture required for a particular crop. Even today, in Punjab, farmers hire labours from Bihar at times of heavy demand period.
3. Time of sowing is very important for high yielding particularly for rainfed sowing for which, all the implements and other necessary accessories have to be kept ready. Any delay in these arrangements received punitive action.
4. Kautilya suggested that for getting good yield of rainfed crop, a rainfall of 16 dronas (one drona = 40 mm to 50 mm) was essential and 4 dronas rainfall is sufficient for rice. It is very interesting to note that during Kautilya's period, rain

gauge was used. It was apparently a circular vessel (20 fingers width, 8 fingers width depth) and the unit to measure rain was adhaka (1 adhaka=12 mm approx.)

5. He also emphasized the optimum distribution of rainfall during crop growing season. One third of the required quantity of rainfall falls both in the commencement and closing months of rainy season (July/Aug; October/Dec) and 2/3 of rainfall in the middle (August/ Sept.; October) is considered as very even distribution of rainfall. This concept is applicable even today for rainfed crop.
6. The crops should be sown as per change in the season. Sali (transplant rice), Virlu (direct sown rice), till (Sesame), millets should be cultivated at the onset of rain. Pulses should be sown in the middle of season. Safflower, mustard, linseed, wheat, barley to be sown later part of the year. It is noted that even today our scientific results have proven that cereals, millets were sown early and oilseeds, wheat, barley which require less water are sown at last or as post rainy season.
7. He also mentioned that rice crop require less labour expense, vegetables are intermediate, and sugarcane requires more attention and expenditure. It is also true even today after 2000 years, the situation has never changed that sugarcane requires heavy labour and expenditure.
8. The crops like cucurbits are well cultivated to banks of rivers. Long-peper, sugarcane and grapes grow well where the soil profile is well charged with water. Vegetables require frequent irrigation, borders of field suited for cultivation of medicinal plants. Even today these practices of growing cucurbit (Watermelon, pumpkin) on river banks, long-peper, sugarcane and grapes in soils with good WHC continue.
9. Some of the bio-control practices suggested by Kautilya has got importance even today. They are:
   (a) Practice of exposing seeds to mist and heat for seven nights to control the diseases. These practices are followed even today also. For example, in wheat to prevent smut diseases, seeds are soaked in water to activate fungal mycelia and then dried the seed under hot sun to kill the fungal.
   (b) Cut ends of sugarcane are dressed with the mixture of honey, ghee and cowdung. Recently it has been proved that honey has antimicrobial property. Ghee could seal off the cut ends and thus, prevent loss of moisture and cowdung facilitated bio-control of potential pathogens.
10. He also suggested that harvesting should be carried out at proper time and nothing should be left in the field not even chaff. The harvested products should be properly processed and stored safely. The crop residues were also removed from fields and used for cattle feed.

B. Trade and Marketing (Economic Policies)

All the industries were categorized into two groups according to their ownership. One key industry group was covered by state and another group by private. It is important to note that this policy resembles today's model mixed economy. The production, distribution and consumption of agricultural produces were well controlled by the king. Agriculture was placed in the category of privately owned industries. The state Government should control and regulate all the economic aspects and evade the influence of market forces and private interests. These practices which were suggested by Kautilya, were followed by Indian farmers for over centuries and these are more sustainable and relevance to scientific agriculture.

C. Astronomy - Prediction of Monsoon Rains; Parashara, Varamihira, Panchanga in Comparison to Modern Methods:

Modern methods of weather forecasting have originated recently. But ancient indigenous knowledge on weather forecasting is unique to our country. India had a glorious scientific and technological tradition. Even now-a-days, it is interesting to note that village astrologers (pandits) are surprisingly right in high percentage of their weather predications.

Meteorology is generally believed to be a new science. It may be new to the West, but not in India, where this science has existed since ancient times. A study of this science was made by our ancient astronomers and astrologers. They use simple rules and costly apparatus are not required. These observations coupled with experience over centuries enhanced to develop meteorology.

The ancient/indigenous method of weather forecasting may be broadly grouped into two categories.

1. **Observational method:** Atmospheric changes, Bio-indicator, Physical changes, Chemical changes, Cloud forms and other sky features
2. **Theoretical methods (or) Astrological factors (or) planetary factors**: Computation of planetary positions and conjunctions of planets
   (a) Study of solar ingress and particular date of months
   (b) Study of Nakshatra Chakras
   (c) Study of Nadi Chakras
   (c) Dashatapa Siddhanta

Alamanacs in Indian Astronomy and Astrology (Panchangs)

According to the Encyclopedia Britannica (1969), "An almanac is a book or table containing a calendar of the days, weeks and months of the year, a register of ecclesiastical festivals and saint's day and a record of various astronomical phenomena, often with weather prognostications and seasonal suggestions for countrymen".

In India, the classical Hindu almanac is known as "Panchang'. This yearly published book is the basic book of the society giving calendrical information on daily basis and is widely used by the people all over India. It is one of the basic book for astrologers to make astrological calculators, casting horoscopes, and for making predictions. For farmers also, it acts as an astrological guide to start any farming activity.

Almanacs in Indian astrology (Panchangs)

According to the Encyclopedia Britannica (1969), "An almanac is a book or table containing a calendar of the days, weeks and months of the year, a register of ecclesiastical festivals, saint's day and a record of various astrological phenomena, often with weather prognostications and seasonal suggestions for countrymen". The book "Almanac" was guide book for farmers since ancient time. In ancient time, farmers always used guide book for smooth agricultural activities.

In India, the classical Hindu almanac is known as 'Panchang'. This yearly published book is the basic book of the society giving calendrical information on daily basis and is widely used by the people all over India. It is one of the basic book for farmers also, it acts as an astrological guide to start any farming activity. In ancient time, farmers used this book to plan cropping system by using the weather predictions and seasonal suggestions.

The other items which are considered for astrological prediction are

1. Rashi or twelve equal parts of the Zodiac belt, therefore, twelve rashis
2. Planets
3. Solar months and solar year
4. Lunar months and lunar year
5. Era

Theoretical Basis of Weather Forecasting in Ancient Literature and Panchangs

According to Varahamihira and other legends, the clouds (or) garbhadharana formation takes place 195 days before their birth (or) delivery (or) garbhaprasava. During this period, clouds were classified as Abartak (Avartak), Sambartak (Samvartak), Pushkara and Drona. If abartak is dominating one year, rain will be received in certain places in that year; if sambartak is dominating in one year, rain will be received in all of the country; the quantity of rain will be very less, if pushkara is dominating in that year; and if drona, abundant rain water will be received in that year. Even today, the cloud classification indicates Circus, Cirrostratus, Cirro cumulus, Altostratus, Altocumulus, Stratocumulus, stratus, Nimbo Stratus, Cumulus and Cumulonimbus. Among these, Nimbostratus and Cumulonimbus give rainfall to the earth.

According to the ruling planet of a year, overall rainfall of that particular year is mentioned in Table 2.1:

Table 2.1: *Ruling planet of a year with the overall rainfall of that particular year*

S.No.	Ruling Planet	Rainfall
1	Sun	Moderate
2	Moon	Very heavy
3	Mars	Scanty
4	Mercury	Good
5	Jupiter	Very good
6	Venus	Good
7	Saturn	Very low (stormy wind)

For predicting the monsoon season and its subsequent influences on weather, all panchang makers consider three different Nadi Siddhantas (Capsular theories) commonly called as Nadi charkas. These are:

1. Dwinadi chakra
2. Trinadi charka
3. Saptanadi charka

Table 2.2: *Arrangement of nakshatras in Saptanadis and its associated effect on weather*

SEVEN NADIS	EFFECT ON WEATHER
Chanda	Bright sunshine, no rainfall
Vata	Sunshine and wind, normal rainfall
Vanhi	Strong hot wind (Westerlies)
Soumya	Normal rainfall
Meera	Very good rainfall
Jala	Abundant rainfall
Amrita	Heavy to very happy rainfall causing flood

Prediction analysis and discussion

The rainfall predictions made in panchangas based on ancient astrological theories are, on an average, better and in some cases at par with the predictions made by Govt. meteorological department through modern techniques and procedures. The yearly truly corrected rainfall predictions were made during 1946-1995 and these were 75, 78, 74 and 75% respectively for different panchangam. The seasonal prediction also mentioned that it was 89% for summer, 55% for rainy, 90% for winter and 78% for overall.

Method of Measurement of Rainfall

The method of measurement of rainfall is mentioned by Varahamihira. A circular vessel with a diameter equal to one (human) arm or the distance measured by the width of 20 (human) fingers and with a depth equal to the distance measured by the width of eight fingers was accepted for measurement of rainfall. When this vessel is filled with rainwater completely, the rainfall should be considered as equal to 50 palas or one adhaka. This method has been explained by the Parashara.

A Model for Forcasting Seasonal Rainfall Recorded in Brhat Samhita

Brhat Samhita mentioned that even in the sixth century AD, Varahamihira, a resident of Malwa (present-day western Madhya Pradesh) faced the problem of uncertainty of monsoon rainfall. The date of onset of such rainfall could not be predicted and so the amount of rainfall during the season was also not predicted. As he was proficient in astrology, he tried to evovle or adapt a technique which was based on that science. This technique lays down that after the occurence of the full-moon day of the month of Jyestha (approximately coinciding with June of Gregorian calendar), the asterism or lunar mansion or nakshatra on the day of first rainfall of that year's rainy season is received should be noted. This lunar mansion provided the basis for the forecast of seasonal rains (Table 2.3). While providing this forecast, it was also important to take into account the area where the first rainfall of the season had occurred. There are twenty-seven such asterisms or lunar mansions in Indian astrology, with each one falling under a particular zodiac sign.

Table 2.3: *Varahamihira's technique for forecasting seasonal rains.*

Sl.No.	Lunar mansion	Zodiac sign		Predicted total seasonal rainfall	
		Sanskrit	English	In ancient units (dronas)	In modern units (cm)
1	Hasta	Kanya	Virgo	16	102.4
2	Purvashadha	Dhanu	Sagittarius	16	102.4
3	Mrugshirsha	Vrushabha	Taurus	16	102.4
4	Chitra	Kanya	Virgo	16	102.4
5	Revati	Meena	Pisces	16	102.4
6	Dhanistha	Makara	Capricorn	16	102.4
7	Satabhisaj	Kumbha	Aquarius	4	25.6
8	Jyeshtha	Vrushchika	Scorpio	4	25.6
9	Swati	Tula	Libra	4	25.6
10	Krittika	Vrushabha	Taurus	10	64.0
11	Shravan	Makara	Capricorn	14	89.6

12	Magha	Simha	Leo	14	89.6
13	Anuradha	Vrushchika	Scorpio	14	89.6
14	Bharani	Mesha	Aries	14	89.6
15	Mula	Dhanu	Sagittarius	14	89.6
16	Purvaphalguni	Simha	Leo	25	160.0
17	Punarvasu	Mithun	Gemini	20	128.0
18	Vishakha	Vrushchika	Scorpio	20	128
19	Uttarashadha	Makara	Capricorn	20	128
20	Ashlesha	Karka	Cancer	13	83.2
21	Uttarbhadrapada	Meena	Pisces	25	160
22	Uttaraphalguni	Kanya	Virgo	25	160.0
23	Rohini	Vrushabha	Taurus	25	160.0
24	Purvabhadrapada	Kumbha	Aquarius	15	96.0
25	Pushya	Karka	Cancer	15	96.0
26	Ashvini	Mesha	Aries	12	76.8
27	Aardra	Mithun	Gemini	18	115.2

1 On the day of the first rainfall of the season 2. 1 drona = 6.4 cm

Krishi-Panchang

The researcher developed the Krishi panchang or Agroalmanac or Agropanchang. It may be defined as basic astro-agricultural guide book/calendar published annually basically region wise, season wise and crop wise, giving calendrical information on various aspects of agricultural and allied activities.

The contents of the proposed Krishi-Panchang can broadly be classified into two major groups as follows:

1. Information which changes every year
   - A. Annual date and Holiday calendar
   - B. Month-wise daily guide for the whole year
   - C. "Rashiphal", i.e., month-wise forecasting of persons havingdifferent zodiac signs.
   - D. Daily/monthly/annual weather forecasting for the particular year
   - E. Crop prospects of that year based on planetary positions
   - F. Season-wise crop strategy based on anticipated weather
2. Information which remains the same irrespective of any particular year
   - A. Theories relating to agricultural and meteorological forecasting
   - B. Auspicious moments for agricultural and allied activities
   - C. Some general agricultural guidance.

Panchang-making

The content and coverage of the proposed Krishi-Panchang, indicate that an editorial board comprising of both qualified astrologers and crop specialists can prepare it, but only qualified astrologers cannot prepare the whole content on their own. While preparing the Panchang, the editorial board members should considered the following important points:

1. The **Krishi-Panchang** is largely meant for the local farming communities, having very low educational status. Hence, it must be in the local colloquial language to facilitate reading and comprehension.
2. Care should be taken to make the Krishi-Panchang easily understandable and clear in its meaning.
3. It should be very comprehensive in its content and coverage with proven predictive information only.
4. It should not contain any astrological details or complexities which would go beyond the understanding capability of our less educated farmers and agriculturists.
5. It should be attractive in colour, and presentation of information should be systematic according to seasons (kharif, rabi, and summer) and crops.
6. It must be low-priced/nominal-priced, within the affordable range of small and marginal farmers.
7. More important is, it must be made available to the farmers and needy persons sufficiently in advance, i.e., at least1-2 months before the start of the agriculture year (july-june).

D. Ancient Soil Classification and Maintenance of Soil Productivity

Soil Classification

In ancient times geographical distribution by Surapala was jangala (arid), anupa (marshy) and samanya (ordinary). It is further divided by colour into black, white, pale, dark, red and yellow by taste into sweet, sour, salty, pungent, bitter and astringent. Samanya land was suitable for all kinds of trees. Rig-veda identified productive and non-productive soils. There were 12 classifications based on soil fertility, irrigation and physical characteristics. These soil classifications are as follows:

1.Urvara or fertile 2.Ushara or barren 3.Maru or desert 4.Aprahata or fallow 5.Shadvala or grassy 6.Pankikala or muddy 7.Jalaprayah or water 8.Kachchaha or land contiguous to water 9.Sharkara or full of pebbles 10. Sharkaravari or sandy 11. Nadimatruka or land water from river 12. Devamatruka or rainfed

Another classification based on crops suitable

(a) Vrdiheyam (rice (rainfed) / corn)

(b) Shaleyam (kamala (wet) rice)

(c) Tilyam (sesamum)
(d) Mashyam (blackgram)
(e) Maudginam (mung bean)

Maintenance of Soil Productivity

Traditional soil management practices are developed from centuries of accumulated knowledge, experience and wisdom over generations. These practices were developed within the framework of local technical possibilities. Ancient farmers mostly depend on crop residues, legumes, manures, and neem for enriching soil fertility.

In Kirishi - parashara, it is mentioned that crops grown without manure will not give yield and emphasized the importance of manures. He also prescribed compost preparation from cow dung. The dried and powdered cow dung is placed in pit for decomposition. The time period for composting is two weeks.

Kautilya mentioned the use of cowdung, animal bones, fishes, milk as manure. Surapala describes the ancient practice of preparing liquid manure (kunapa) prepared by boiling a mixture of animal excreta, bone marrow, flesh, dead fish in an iron pot and then add it to sesame oil cake, honey and ghee. This is clearly noted that present day Panchakavya is prepared in the same method and used in all crops.

Liquid manure (Kunapa): Preparation of kunapa involves boiling flesh, fat, and marrowof animals such as pig, fish, sheep or goats in water, placing it in earthen pot, and adding milk, black gram boiled in honey, powders of sesame oil cake, decoction of pulses, ghee and hot water. There is no fixed proportion of ingredients. The pot is put in a warm place for two weeks. This fermented liquid manure is known as kunapa.

Green Manures

In Rajasthan: *Prosopis cineraria* - this brings up moisture and nutrients from the underground and leaves are used as green manure.

In North India: A traditional weed *Kochia indica* is used as green manure.

Ancient farmers adopted inter cropping and crop rotation to restore soil fertility. Mixed or inter cropping with legumes in cereal and oil seed cultivation were widely practices. All these practices adopted in ancient time are now being recommended today under organic farming system.

E. Water Harvesting and Irrigation Developments During Different Periods – Water Storage – Distribution and Relevance to Modern Agriculture.

The continuous supply of water for irrigation whether from canal, well, pond or lake is realized as the most important factor for agriculture in ancient period. It should be preserved by all sorts of efforts for the benefit of agriculture. The different principles of irrigation adopted in ancient period are:

1. Construction of large mud embankment on a stone foundation for diverting flood water from the river.
2. Building of small tanks.
3. It is also indicated that severe penalty was imposed when water is letout other than sluice gate.
4. Extensive tank irrigation systems were adopted in Sri Lanka and later in South India. In Srilanka, ancient kings practiced that not even a drop of rainfall should go to sea without benefiting man.
5. The topography of Telengana region of Andhra Pradesh and Karnataka is ideally suited for the construction of tanks.
6. Every farmer emphazises the efficient and skilful use of water.
7. It is also suggested that preference of the use of water should be in the order of food crop, vegetables and flowers.
8. For the mainteance of tanks, a committee of villages called `eri- variyam' was appointed. The committee ensured repairs and de-silting tanks and distribution of irrigation from wells.

Bullocks were used to draw water from wells for irrigation. Persian well which was developed in North India, was used for drawing water from wells.

History of Irrigation Development in India

Sl.No.	Period		Irrigation Development
1	Ancient Period	2500 - 1000 BC	People started to settle near the banks of river/ tanks for the purpose of getting water for drinking and irrigation.
2	Chalcolithic	3000 - 1700 BC	Practice of irrigation to crops was evolved.
3	Vedic period	1600 - 1500 BC	People engaged craftsman to dig channels from rivers to their fields. Well irrigation through kuccha and puccha wells and were practiced
4	Pandyas / Cholal chera's Period	(1st Century 300 AD)	Irrigated rice cultivation started during this period. Dams and Tanks were constructed for irrigation purpose.
5	Medieval period	(1200 - 1700 AD)	Irrigated agriculture was developed during Mughal period. Canals, Dams and Tanks were constructed. For example 1. Construction of western Yamuna canal 2. Constructions of Anantaraja sagar.

Methods of Conserving Rain Water

In ancient days itself, people, especially Indians, know the methods of conservation of rain water. There are instances that there was very good system of water management even during Harappan period. During Independence period, the people considered the water as part of the nature which is essential for their survival and used to manage water resources very precisely. This could be proven from the rain water harvesting structures in the low rainfall areas of Rajasthan, harvesting springs in mountainous region and hilly areas and percolation ponds and tanks in southern India.

In Tamil Nadu, the ancient people used to store rainwater in public, placed separately one for drinking purposes and another for bathing and other domestic purposes and called them as Ooranies. They also constructed percolation tanks or ponds, for the purpose of recharging irrigation or domestic wells. They used to periodically clean the water ways so as to get clean water throughout the year. These are instances in the history that people constructed crude rubble bunds across river courses either for diversion of water or for augmenting the ground water. The various methods of rainwater harvesting are categorized under two groups – (A) Traditional methods and (B) Modern methods.

(A) Traditional Methods

Traditional rainwater harvesting, which is still followed in rural areas, was done in surface storage bodies like lakes, ponds, irrigation tanks, temple tanks etc. In urban areas, due to lack of open spaces, rainwater is harvested as ground water,

Hence harvesting of rain water in such places will depend very much on the nature of the soil viz., clayey, sandy etc. The various kinds of traditional rainwater harvesting methods are listed below.

(i) **Bamboo Method of Rainwater Harvesting:** In Meghalaya (one of the seven northeastern states in India), an indigenous system of irrigation by using bamboo pipes to irrigate plantations is widely practiced. It is so nicely constructed that about 18-20 litres of water entering the bamboo pipe system per minute gets transported over several hundred metres with in the pipes and finally it is reduced to 20-80 drops per minute at the site of the plant. The tribal farmers of Khasi and Jaintia hills are using this 200-year-old system.

The bamboo drip irrigation system is mostly used to irrigate the black pepper or betel leaf crops planted in arecanut orchards or in mixed orchards. Bamboo pipes are used to divert perennial springs rainfall on the hilltops to the lower part by gravity. The channel sections which is made of bamboo, divert and supply water to the plot site where it is distributed without leakage into branches which are again made and laid out with different forms of bamboo pipes. The intake pipe position's manipulation also controls the flow of water into the lateral pipes. Reduction of channel sections and diversion units are used at the last stage of water application. The last section of the channel enables the water to be dropped near the roots of the plant.

(ii) **Kunds of Thar Desert:** The Thar Desert villagers had developed an indigenous system of rainwater harvesting in the sandy tracts popularly called as kunds or kundis. Kund is the local name given to a covered underground tank. It was developed primarily for tackling drinking water problems constructed usually with local materials or cement. Kunds were more prevalent in the western arid regions of Rajasthan, and in areas where the limited groundwater availability is moderate to highly saline. Groundwater in Barmer, for example, in nearly 76 per cent of the district's area, has total dissolved salts (TDS) ranging from 1,500-10,000 parts per million (ppm). Under such situations, kunds provide convenient, clean and sweet water for drinking. Kunds were owned by communities or privately, with the rich having one or more kunds of their own. Community kunds were built through village cooperation or by a rich man for the entire community.

(iii) Traditional Rain water harvesting: The traditional rainwater harvesting methods in North India methods are Tanka, Nada, Nadi, Talai, Talab, Khadin Sar, Sagar and Samend. Depending upon rainfall intensity, topography of area and type of soil, the water harvesting methods are different from region to region.

(iv) Tanka: It is one of the oldest, common and relatively hygienic method of water storage. It is constructed on farm, country yard and fort. It is normally circular or square shaped. Dimension is 2 m in diameter, 3 m in deep and capacity is 10000 lit. It is constructed on sloping land to arrest runoff water in the farm. However, the construction is also made in houses on an elevated place to avoid entry of dirty water in to it.

(v) Talai: Similar to Tanka, but it is still deeper (2-3cm depth). Special attention was made for selection of location such that there is adequate flow of rain water into Talai. Care is also taken so that loose soil does not flow along with water runoff.

(vi) Nada: In this method, low lying areas in between hillocks is dug out as pit and provided embankment to catch rain water from these hillocks. The catchment area of Nada is 5 to 10 ha. It is normally constructed on rangeland, barren land pastureland and agriculture field. It provides short-term storage of rainwater which is mainly used for animals.

(vii) Nadi: Compared to Nada. the Nadi is bigger in size. A village or group of Villages uses the collected runoff water in the Nadi. Depth is 6-8m, catchment area 10-150 ha. In the Nadi, water is available for whole of the year which provides shelter for many wild animals and birds.

(viii) Talab: It is comparetively shallow and spread over to more area as compared to Nadi. It is generally constructed in rangeland. The catchment area of Talab is 480 ha.

(ix) Khadin: Accumulation of runoff water in between hillocks is known as Khadin. Khadin means cultivation crops. The khadin water is generally used for crop cultivation and animals.

(x) Sar, Sagar and Samand: It is used to collect rainwater for irrigation purpose. This structure provides excellent source of reservoir and also tourist spot.

(B) Modern Methods

The Modern methods of rainwater harvesting are categorised under two, they are Artifical Recharging and Rain Water Harvesting. The former is classified into Absorption Pit Method, Absorption Well Method, Well cum Bore Method and Recharge trench cum injection well. The later is grouped into Individual Houses and Grouped Houses which are further categorized into Percolation Pit Method, Bore Well with Settlement Tank, Open Well Method with filter bed Sump and percolation Pit with Bore Method.

Practices of irrigation and rainwater harvesting adopted in ancient period were more relevant in Indian agriculture today.

- Water is life and kingpin of successful agriculture.
- Emphasis to be given for water conservation for efficient use for agriculture and domestic needs.
- Sustainability of irrigation system was very much emphasized where the tanks, wells or canals must ensure continuous supply of water.
- Avoidance of salinization/alkalinization and water logging problems which become features of canal irrigation system of today.

F. Plant Protection in Ancient Period–ITK–Harvesting–Threshing and Storage

In the past, when Indians were gathering knowledge on the prediction of rainfall, management of agriculture, farm activities, harvesting, and storage, nothing was known about plant protection. The only methods to protect the crop were prayers and mantras. It was believed that the crop will be protected if the mantra was tied to the crop which were written with red lac-dye. But it cannot be said that the people of that time were unaware of insects and other pests and their damage. Some of the pests (in Sanskrit) affecting crops were gandhi, Shankhi, Pandarmundi, dhuli, and shringari. It is interesting that gandhi (offensive odour) is what is called today the gandhi bug (*Leptocorisa varicornis* F.); shankhi must be a snail (*Pila sp.*); and pandarmundi means white head which is the typical symptom of the attack of rice stem borer (*Tryporyza incertulus* Walker). It is certain that they knew the rice stem borer and its symptom of attack. Dhuli means powder and it is certain that this word must have been used for powdery mildew of barley and wheat. The word "shringari" in Sanskrit indicates something adorned with red colour and it is possible that the term was used for rust diseases.

Besides these pests, goats, rats, pigs, deers, wild boars, sparrows and parrots were mentioned as destroyers of crops. In fact when the damage to crops due to different pests reached the economic injury level, they might have started thinking about plant protection and diverted their efforts to develop protection technology. It is significant that people at that time considered that plants and human beings have similar physiology. Therefore, they classified the diseases of plants into two groups: (1) internal; and (2) external. The internal diseases were those which were caused by "vata", "pitta", and "kafa" and external diseases were those which were caused by insects, birds, and weather. These categories can be attributed today to fungi, bacteria, viruses, and nematodes-internal diseases; and insects, non-insect pests, frost, waterlogging, and drought-external diseases.

Table 2.5: *Information contained in Surapala's Vrikshayurveda , related to kinds of internal disorders observed in trees, causes and symptoms attributed, and remedies*

Cause given	Symptoms	Cause elaborated	Possible causes2
Vata	Trunk slender and crooked; knots on trunk or leaves; hard fruits (less juicy and sweet); gradual defoliation; flower and fruit drop; generally yellowing of leaves and fruits.	Arid land on account of excessive supply of dry and pungent matters.	Underground mechanical barrier, leaf-galling insects; root-infecting fungi or ematodes;viruses; saline or alkaline soils.
Pitta	Leaf yellowing; premature drop; decay of flowers and fruits	Occur at the end of summer if trees are excessively watered with bitter, sour,salty and strong materials	Viral disease; salinity in irrigation water; predisposal to blossom blight; fruit decays due to fungal/bacterial infections
Kapha	Fruit-bearing delayed and fruits are tasteless and ripen prematurely; oozing without wounds	Appear in winter and spring if trees are excessively watered with sweet, oily, sour or cold materials	Fungal gummosis/ rot;nutrient deficiencies or toxicities; excessive watering

Today integrated pest management (IPM) is considered a recent approach for plant protection but the so called recent approach was conceptualized and practised centuries ago in India. Some of the practices adopted in those days are given below.

1. **Seed treatment:** Seed treatment which is considered an important component of IPM to ensure better germination was given a lot of importance in ancient times. The seed was treated with milk, mustard, sesame-ash, and cowdung for better germination and protection against insect pests.
2. **Fumigation:** Fumigation methods were not as developed in those days as they are today but the concept of fumigation was prevalent. For e.g., diseases of cucurbits were controlled by smoking the bones of cow and dog mixed with the excreta of cat.
3. **Field application:** The plant protection materials at present era were not developed during that time; sprinkling of liquid suspension and hand-dusting of various products were used. For the control of insect pests several ancient recommendations are available. Some of them are as follows:

   (a) Insects infesting trees can be removed by smoking a mixture of white mustard, black peper, asafoetida, vidanga (Embelia ribes), vaca (Zingiberzerumbet), and water mixed with beef, horn of a buffalo, flesh of pigeon, and the powder of bhillata (Semecarpus anacardium).

(b) Insects infesting creepers can be controlled by sprinkling water mixed with oilcake.

(c) Leaf-eating insects can be destroyed by dusting cowdung-ash and brick-dust.

(d) Trees are watered with cold water for 7 days to remove insects from the roots and branches.

(e) A wound caused by insects is healed if sprinkled with milk after being anointed with mixture of vidanga, sesame, cow's urine, ghee, and mustard.

A new term, eco-friendly pesticides, has been coined recently. In IPMmore emphasis is laid on this term and botanicals are being used instead of chemical pesticides. In fact this is not new. Few years earlier, several botanicals and other materials having biocidal properties, were identified and suggested by Surapala to control plant diseases. The famous "panchamula" (roots of five plants) which was commonly used at that time has antifungal, antiviral, antibacterial, and antifeeding properties. Likewise, mustard had been used for all kinds of diseases caused by "kafa". We now know that mustard causes antibiosis in insects; in addition it is antifungal and has nematicidal activity.

Some Important Products Used in Pest Management During Ancient and Medieval Periods in India

Material	Author/Period	Properties
Root of vasika (*Justicia adhatodaa*)	Varahamihira (505-587 AD)	Soothing effect, insecticidal, antifungal, antibacterial, anthelmintic.
Branches and leaves of atimuktaka(*Hiptage enghalensis*)	Varahamihira (505-587 AD)	Leaf juice insecticidal; bark contains glucoside (hiptagin) and tannins
Mustard (*Sinapis alba* = *Brassica alba*)	Surapala (1000AD)	Insect antixenosis and antibiosis; acaricidal; nematicidal; antifungal
Bidanga (vidanga) (*Embelia ribes*)	Surapala (1000AD) Someshwara Deva (1126 AD)	Anthemintic; antibacterial insecticidal
Ash	Someshwara Deva (1126 AD)	Dessicates insect eggs on seed; speeds up germination by softening seed coat through mild alkalinity; provides micronutrients
Sesame (*Sesamum indicum*)	Surapala (1000 AD)	Allelopathic to rice; insect repellent; insecticidal properties
Mahua (*Madhuca spp.*)	Surapala (1000 AD)	Insecticidal oil; piscicidal; antibacterial

Material	Author/Period	Properties
Kusta (costus) (*Saussurea lappa*)	. Surapala (1000 AD) .	Insecticidal (repellents, anti-feedant); antiseptic
Bhillata (Bhallataka) (*Semecarpus anarcardium)*	Surapala (1000 AD)	Insecticidal; antiseptic' termite-repellent;mildew moth-proofing of cloth anthelmintic; antibacterial
Cotton(*Gossypium spp)* seed oil	.	

Materials and Practices that Need our Early Attention

(a) Milk and milk products: Milk and ghee have been used for centuries. Even buttermilk was found useful. About 40% of total aminoacids in milk are glutamate, leucine, and proline. Milk contains plant growth promoters. A recent study reported that milk sprays induced systemically acquired resistance in chilli against leaf curl which is a viral disease. Milk (10% aqueous suspension) also has been effectively used for controlling powdery mildews. Besides, milk has excellent sticker-spreader properties.

The amino acid, proline, has been found to systemically induce resistance in plants. It stimulates production of antimicrobial phenolics. High amounts of endogenous proline increase contents of cytokinin and auxins. Besides milk, proline is present in the connective tissues of animals including fish.

(b) Application of cowdung: Use of cowdung for dressing seeds, plastering cut ends of vegetatively propagating units such as sugarcane setts, dressing wounds, sprinkling diluted suspension on plants, and applying to soil has been indicated since the time of Kautilya (c.300 BC). Indian farmers continue to use cowdung in various ways, but the agricultural scientists have ignored its use for other purposes except manure.

Briefly speaking cowdung from the cattleshed is a mixture of dung and urine, generally in a ratio of 3:1. Cowdung consists of crude fiber, crude protein, and materials that can be obtained in nitrogen-free extracts and ether extracts. Cellulose along with lignin makes up most of the crude fiber; hemicellulose and pentosans (poly saccharides based on pentose sugars) are also present. Micronutrients too are present in cowdung. The urine portion of cowdung contains nitrogen, potash, and sulfur and only traces of phosphorus. The nitrogenous compounds excreted in fecal matter consist in part undigested or unabsorbed food nitrogen and in part another fraction called metabolic nitrogen. The metabolic fraction comprises substances originating in the body such as residues of the bile and other digestive juices, epithelial cells from the alimentary tract, and the bacterial residues. In short, fecal residues comprise undigested fiber, debris from sloughed-off intestinal epithelium, some excreted products derived from bile (eg. pigments), intestinal bacteria, and mucus. There are more than 60 species of bacteria and over 100 species of protozoa encountered in the rumen of a cow. A majority of the bacteria are cellulose, hemicellulose, and pectin fermenters. The bile constituents are bile salts, bile acids, and bile pigments. Bile salts contains hydrophilic coat to otherwise hydrophobic droplets, thus acting as emulsifying agents. No bile salt is supposed to be present in the dung because these are reabsorbed through the intestine and are put back in the bile. Bile salts have antiseptic properties. Two chief bile pigments are bilirubin (reddish / golden yellow) and biliverdin (green). It is the biliverdin ($C_{33}H_{36}N_4O_8$) which is chiefly present in herbivorous animals and gives greenish color to the dung.

Table 2.6: *Materials suggested by Surapala to control tree disorders and their currently known properties.*

Materials (Plant species)	Properties
Acorus calamus L.	Antibacterial
Brassica alba (L.) *Rabenh*/ *Sinapis alba* L.(white mustard)	Insect antixenosis; antifungal; acaricidal; nematicidal; glucosinolate sinalbin "anti-insect" and "anti-nematode" allyl isothiosinate antifungal
Curcuma longa Koenig non L. *Curcuma domestica* Val. (turmeric)	Antioxidative curcuminoids; antimicrobial
Embelia ribes Burm. F	Anthelmintic; antibacterial; insecticidal (embelinbenzoquinone)
Emblica officinalis Gaertn. (triphala)	Anthelmintic with other two species of triphala.
Ficus benghalensis L. (Banyan)	Latex with good sealing property; tannin
Ficus glomerata Roxb.	Latex; bark 14% tannin; some Ficus spp. are antibacterial.
Piper nigrum L. (black pepper)	Oleoresin antibacterial/antifungal;alkaloid piperin is insecticidal.
Sesamum indicum L. (Sesame)	Insecticidal and repellent; oil synergistic to pyrethrums; antioxidative lignins in seed; 17% protein; 800 mg per 100g calcium, phosphorous, and potassium; 14% iron (ash) - highest.
Solanum indicum L.	Fruits/leaves antifungal /antibacterial; glyco- alkaloid solasonine present.
Materials (Animal products and other materials)	
Ash	Particles hygroscopic; absorb moisture from insect eggs and spores; interfere with insect feeding; ash potassium interacts with surface fats.
Cowdung	With urine it is antiseptic; rich in bacteria which compete with pathogens; good medium for biocontrol agents; beneficial to Rhizobium and Azotobacter.
Fish meal	Rich in protein; releases aminoacids including proline.
Ghee	Same as animal fat
Honey	Antimicrobial; protects wounds in plants / animals; proline present; honeybee peptide apidaecin is antibacterial.
Liquid manure (kunapa)	Effects would include: healthy crop/tree; crop tolerance to abiotic stresses such as frost, heat, etc. as well as to insect pests and disease; high yields; high quality produce.

G. Harvesting and Threshing

Kautilya Arthasasthra states "Grains and other crops shall be collected as often as they are harvested. The threshing of different fields shall be in close proximity. It is mentioned in Sangam literature that paddy was removed from the stalks by beating them on ground or by making the bullocks to tread on them. After proper cleaning, paddy was collected, measured and stored in proper places. Sickles and swords were used for harvesting of millet. Sticks were used to thresh the blackgram. Threshing and cleaning were mostly done by the women. A common vessel for measuring grain which was referred as "ambanam"

Festivals were celebrated before the starting of harvest and also during the time of harvest. Parashara mentioned a levelled threshing "pit" and installation of threshing pillar called "medhi". For this pillar, wood was obtained from a tree that produces milky sap, preferably by silk cotton, *Ficus bengalensis*, *F. glomerata*.

Measurement

'Adhaka' is wooden vessel made of mango, punnaga (*Callophyllum inophyllum*). It is used to measure grains which is equal to approx. 11 oz or 3.5 Kg.

I. Gardening in Ancient and Medieval Period (Arbori Horticulture and Orchards)

Man cannot be separable from nature. Since prehistoric times, the Indian people have close relationship with nature, particularly plants and flowers. The tree worship has been a traditional ritual and faith in India through the ages. Trees and flowers have been sanctified in Indian mythology, history, art and socio-religious culture. The pre-historic and proto-historic man in the Indus Valley had great reverence for trees and worshipped them during the chalcolithic period. Tree was worshipped in its natural form and as tree spirit personified as human attributes in Mohenjo-daro and Harappa. The character of the trees has been depicted in a number of seals, potteries, potsherds and some rock paintings as archeological evidences of the Mohenjodaro and Harappa period (2500-1750 BC). A few trees, such as pipal or asvattha (*Ficus religosa*), neem (*Azadirachta indica*), katha or khadira (*Acacia catechu*) and jhand or sami (*Prosopis cineraria*) were held sacred by the ancient people of the Indus Valley. During the Indus Valley Civilization, the belief was that trees were symbolic of gods and goddesses, which lived in them as vriksha devata or vriksha devi. In ancient India, trees were considered to be divine and spiritual as the tree knowledge (brahma tarn), the tree of life (Uivan tarn) and also as medicinal tree (rogu tarn).

Mauryan Period

After the rise of the Mauryas during 4th to 3rd century BC, there has been vast secular literature and texts, both Vedic and post-Vedic, like Vedas, Brahamanas, Aranyakas, Upanishadas, Sutras, Smritis, Mahakavyas, Puranas, Buddhists texts (Jataka) and Jain literature (Sutras). The sages of the Upanishadas have described the Cosmic Tree rooted in the Brahman, the ultimate, whose branches are space, wind, fire, water and earth. The cosmic tree is the World Mother, the Goddess of nature, which nourishes all life. Kalpavrksa (wish-fulfilling tree) and kalpalata (wish-fulfilling creeper) are mythological tree and creeper, not mentioned in the Vedic literature, which have been a part of folk cult in Hindu mythology. Kalpavrksa is mentioned in Ramayana, Mahabharata, Jatakas, Divyavadana and the Jain Sutras. In Brahamanical religion, vata (*Ficus benghalensis*) was identified with Shiva, asvattha (*Ficus religiosa*) with Vishnu, lotus with Surya (Sun) and nine leaves of nine trees (navapatrika) with nine different aspects of Durga.

There are different descriptions of trees in the Rigveda (3700–2000 BC), the Ramayana (1200–1000 BC), the Mahabharata as well as other literature by Shudraka (100 BC), Kalidasa (c. 57 BC), Ashvaghosha (100 AD), Vatsyayana (300–400 AD) and Sarangdhara (1300 AD). The art of gardening and kinds of gardens were described by Sarangdhara and Vatsyayana, respectively.

In the Ramayana, Ashokavana or Panchavati, in which Sita was held captive, is mentioned. Ashoka trees (*Saraca asoca*) were predominant in this garden. In the Panchavati, five trees were planted, asvattha (*Ficus religiosa*) on the east side, bilva (*Aegle marmelos*) on the north, the banyan (*Ficus benghalensis*) on the west, amla (*Emblica officinalis*) on the south and the ashoka (*Saraca asoca*) on the southeast. A description of the layout of gardens and parks and artificial lakes in the city of "Indraprastha" is given in the Sabha-Parva of the Mahabharata. Several trees, like *Saraca asoca, Terminalia arjuna, Mesua ferrea, Ficus benghalensis, F. religosa, Michelia champaka, Butea monosperma* and *Cassia fistula*, have been mentioned in the Ramayana. Almost all of them also have been described in the Mahabharat. It is said that Lord Buddha was born under sal tree in a garden. The bodhi tree, under which the Buddha attained nirvana, is sacred to Buddhists. The trees and plants mentioned in Buddhist texts include asvattha (*Ficus religiosa*), banyan (*Ficus benghalensis*), udumbara (*Ficus glomerata*), patali (*Bignonia suaveolens*), sala (*Shorea robusta*) and sirisa (*Acacia sarisa*). The planting of roadside avenue trees (margeshuvriksha) was an important contribution of the king Ashoka (233 BC).

Vedic Times

The trees and the plants mentioned in the Vedic times were, soma (*Sarcostemma acidum*), pipal/asvanha (*Ficus religiosa*), sami (*Prosopis ceneraria*), banyan (*Ficus benghalensis*), udumbara (*Ficus glome rata*), bilva (*Agele marmelos*); khadira (*Acacia catechu*), neem (*Azadirachta indica*), palasa/plpksa (*Butea monosperma*), tulsi (*Ocimum sanctum*) and lotus (*Nelumbo nucifera*). The other trees and plants of the Vedic and post-Vedic period include salmalilsilk cotton tree (*Bombax ceiba*), coconut (*Cocos nucifera*), rudraksha (*Elacocarpus sphaericus*), snuhi (*Euphorbia neriifolia*), madhavi lata (*Hiptage madablota*), amalka (*Emblica officinalis; syn. Phyllanthus emblica*), mango (*Mangifera indica*), banana, plantain / kadali or rambha (*Musa paradisiaca*), ber / vadari (*Zizyphus mauritina*), sala/shal (*Shorea robusta*), asoka (*Saraca asoca*), kadamba (*Anthocephalus cadamba*), bahira (*Terminalia belli rica*), arjuna (*Terminalia arjuna*), nagavalli, tambula or paan (*Piper betle*), nalaka (*Arundo *donax*), jivaka (*Putranjiva roxburghii*), mandara (*Erythrina variegata*), tili/tila (*Sesamum indicum*), amarphal (*Monstera deliciosa*), *Ficus krishnae*, gaduchi (*Cocculus cordifolius*), ketaki (*Pandanus odoratissimus*), imli/tintrini (*Tamarindus indica*), aparijata (*Nycanthes arbortristis*) and tinduku (*Diospyros peregrina*).

Mughal Period

With the renaissance of gardening in India by the Mughal rulers, many plant species were brought by them from Persia and Central Asia where various herbaceous and bulbous flowers were already under cultivation. Many of these have been mentioned in autobiographies and other books written during those days. Besides, in Mughal paintings also we find illustrations of many flowers. These have also been used to illustrate the borders of those paintings. In the book, "Bagh-I- Wafa".

During the 16th and 17th centuries AD, Mughal gardens were developed in Agra, Delhi, Pinjore (near Shimla) , Srinagar, Kashmir and a few other places. The most important Mughal gardens are the Taj Mahal Garden, Agra; Shalimar and Nishat Gardens, Srinagar; PinjoreGardens, Pinjore and the garden at Hamayun's tomb, Delhi. The rose was introduced into our country via the port of Bussorah by Babur in around 1526. Jehangir and Nurjehan were ardent lovers of the rose and encouraged rose growing in gardens. The most important plants introduced in Kashmir from Persia by the Mughal ruler, Jehangir in 1619 when he laid out the famous Shalimar Bagh in Srinagar, were the majestic Chinar tree (Platanus orientalis), the cypress (Cupresus sempervirens) and the weeping willow (Salix babylonica), and flowers like rose, narcissus, daffodil, iris, lilies, tulip and carnation

(a) The concept of developing a garden in an enclosed space was introduced by the Mughals in India during 16th and 17th centuries.

(b) Babur is credited with the introduction of scented Persian rose in India.

(c) Akbar the Great (1556-1605), the Mogul emperor of India was the garden lover.

(d) Abu-i- Fazi provided a list of 21 fragrant flowering plants along with flower colour and season of flowering in Ain-i-Akbari.

(e) He also gave another list of 29 plants with flowers notable for their beauty.

(f) From the Tuzuk-i-Jahangiri it appears that Jahangir was familiar with nearly all important fragrant palnts of India like Michelia champaca, Pendanus odoratissimum, Mimusops elengi, Jasminum officinale. Mughal gardens were developed in Agra, Delhi, Pinjore (near Shimala), Srinagar, Karhmir and a few places during the 16th and 17thcenturies AD.

(g) During the Mughal period (16thand 17th centuries AD) and the British period (18th and 19th centuries) several ornamental plants were introduced into India. Indian native flora has made significant contributions to the gardens of the world and also to the improvement of a few flowers like Rhododendrons and orchids.

European Period

- Missionary priests, Portuguese, Englishmen, amateur and professional gardeners from Europe, Asia and Africa, introduced a large number of plants in Indian gardens. One of the important missionaries who introduced a number of exotic plants was Dr Firminger, an Englishman, who wrote a book on gardening titled with 'Firminger's Manual of Gardening in India' in 1863 which is giving descriptions of various species of flowers. This book is an authoritative reference book on ornamental flowering plants even today.
- Several Government botanical gardens were established by the British rulers during 18th and 19^{th}centuries in various parts of India, such as Lalbagh Botanical Garden, Bangalore (1760); the Government Botanic Garden, Saharanpur (1779); the Indian Botanic Garden, Sipbur Calcatta (1783); the Lloyd Botanic Garden, Darjeeling (1878) and the Government Botanic Garden, Oatacamud (1884), where number of economic indigenous and exotic plants were introduced and maintained. Among the noteworthy introductions of that period are the mahogany (*Swietenia mahogany*) from Jamaica in 1795 and the Giant Amazon lily, Victoria regia, into Sibpur garden followed by *Grevillea robusta* and *Araucaria excelsa* in 1857 and *Amherstia nobilis* in 1859 in the Lalbagh Botanical Garden, Bangalore. A few important and rare flowers of India are *Agapetes auriculata. Corydalis govaniana. Dendrobium chrysanthum, D. nobile. Geranium wallichianum, Katherinea ampla, Meconopsis aculeate. Notholirion thomsonianum, Nepenthes khasiana, Rhododendron macabeanum, R. hodgsonii. and R. thomsonii.*

Ancient Tree Art and Craft Culture

Tree motifs have been found in the art of Indus Valley, Mauryan ring stones, and gateways and railings of stupas at Bharhut, Bodhgaya, Sanchi, Amaravati and Nagajunakonda, Mauryan relief sculpture. Old architecture and sculptures of Mathura (Kanishka period, AD 78–101) and Ajanta frescoes (AD 100–600) also indicate the importance of plants and flowers in ancient India. The relationship of trees with the Brahamanical and Buddhist Gods and Goddesses and Jain Tirthankaras in Indian art still persist since from the 151 century AD to 1200 AD. Trees and flowers have been also found in ancient coins at the pre-Mauryan site, Sugh, Taxila, Ayodhya during Mitra Kings, Kausambi and Mathura and also of the Andhra dynasty and Pandyan territory. The ancient Sanskrit and other literature and texts, mythological epics and legends, paintings, cave murals and frescoes, sculptures, architecture, folklores and tribal arts and crafts provide evidences of the kind of plants, trees and flowers growing in the forests and gardens. The science of plant life, (Vrikshayurveda) and arbori-horticulture, and usefulness of forests and gardens were well known in ancient India. They were concerned with the conservation of trees and biodiversity in nature and ecological balance in environment. Their concept of identifying trees with Gods and Goddesses, and threats and punishments against the destruction of useful trees helped to save the trees and flora, which is a remarkable contribution of our ancient people.

Native Indian Ornamental Plants Trees

Albizzia lebbeck, Alstonia scholaris, Anthocephalus cadamba, Azadirachta indica, Bauhinia variegata, Bombax malabaricum, Butea monosperma, Cassia fistula, C. nodosa, C. siamea, Erythriana indica var. parcellii, E. Variegata var. orientalis, Ficus religiosa, F. benghalensis, F. benjamina, F. elastica, F. infectoria, F. retusa, Lagerstroemia speciosa, Michelia champaka, Milletia ovalifolia, Millingtonia hortensis, Mimpusops elengi, Polyalthia longifolia, Pongamia pinnata, Pterospermum acerifolium, Putranjiva roxburghii, Rhododendron arboreum, Saraca asoca, Tecomella undulata, Terminalia arjuna, Thespesia populnea.

Shrubs

Barleria cristata, Buddleia asiatica, Daedalacanthus nervosus, Dombeya spectabilis, Holmskioldia sanguinea, Jasminum sambac, J. grandiflorum, J. humile, J. officinale, Nyctanthes arbortristis, Russellia juncea, Spermadictyon suaveolens.

Climbers

Clematis paniculata, Clitoria ternatea, Ficus repens, Hiptage benghalensis, Jasminum grandiflorum, J. humile, J. officinale, Porana paniculata, Thunbergia grandiflorum.

Seasonal Flowers

Amaranthus spp., Celosia, Gomphrena, Lady's lace (Pimpinella monoica), Torenia.

Bulbous Flowers

Allium giganteum, Canna, Gloriosa superba, Iris spp., Tulip spp., Orchids (Aerides, Anoectochilus, Arachnis, Arundina, Calanthe, Coelogyne, Cymbidium, Dendrobium, Paphiopedilum).

Other Plants

Lotus *(Nelumbo nucifera), Nymphaea spp. (N. pubescens, N. rubra, N. stellata)*

Indoor Foliage Plants

Aglaonema, Asplenium nidus, Begonia rex, Coleus blumei, Pilea cadieri, Pteris cretica var. crispate.

Indoor Flowering Plants

Crossandra infundibuliformis, Gynura aurantiiaca, Kalanchoe blossfeldiana.

Floriculture in Ancient India

Plants were also used in personal adornment and beautification of the home. Of the climbers, Madhavilata (*Hiptage madhablata*) received frequent mention in Kalidasa's play

(5th century) and among sweet scented shrubs the mask-mallow (*Hibiscus abelmoschus*) and the garland flower (*Hedichium coronarium*). Description of flowers and gardens had been presented in ancient Sanskrit classics like Rig Veda (3000-2000 B.C), Ramayana (1200-1000 B.C) and Mahabharata (500 B.C). Other Sanskrit books of early days wirtten by Shudraka (100 B.C.), (Asvaghosha (100 A.D) and Sarnghara (1283-1301 A.D) also mentioned about flowers and gardens.

Among the flowers the sacred lotus (*Nelumbo mucifera*) was the most important and numerous references to it occur in Sanskrit literature. Int he days of Moahendjadaro, lotus blossoms were wreathed over the head of Sun-God. The Hindus were so fond of ornamental plants that some of them were actually worshipped.

H. Medicinal Plants and their Relevance in Today

- Medicinal and aromatic plants have been used for a long time for their medicinal properties.
- About 2000 native plants species have medicinal properties and 1300 species are known for their aroma and flavour.
- The Indian systems of medicines, popularly known as Ayurveda, unani and sidha drugs are of great demand in the country.
- There is already spurt in demand of plant based drugs and of late may such native species of medicinal values are being brought under systematic cultivation.
- Opium poppy is perhaps the exotic plant brought under cultivation in the later part of 16th century through introduction from West Asia. It is one of the cultivated crops where the selection pressure has played a major role in developing new varieties and land races.
- Cinchona has been introduced in India mainly to fight the menace of malaria and diahorrea respectively, while belladonna, ergot, henbane, pyrethrum,and toxglove were introduced in India during second world war to meet the demand of raw materials for vital drugs.
- Out of these, about 65 plants have large and consistent demand in world trade.
- India however produces only limited quantities of these materials.
- In terms of market share in production value, India holds only the 6th place with a mere 7% share.
- At present, about 90% collection of medicinal plants is from the forests and many useful plant species are endangered or threatened since 70% of the plant collections involve destructive harvesting.
- In India, more than 15 lakh practitioners use medicinal plants in preventive and curative application.

Two ancient systems of medicine in India are available, the Ayurveda prevalent in the North and the Siddha that flourished in the South. Instead of providing the name of any one individual as the founder of either system, our ancients wisely attributed their origin to the Creator. In the case of Ayurveda it was Brahma, the Creator of the Universe, who taught the science to Prajapati, he to Aswini Devatas and they, in their turn, to Atreya etc. So this tradition is called the Brahma or Arsha Sampradaya (the tradition of Rishis). The inference to be drawn from these traditions is that, there is no exact point of time to which the beginning of these systems could be traced. They are eternal, without a beginning or end; they began with man.

Relevance of Medicinal Plants Today

The World Health Organisation (WHO) reported that 80 % of the population of developing countries still apply traditional medicines, mostly plant drugs, for their primary health care needs. Also, modern pharmacopoeia develops at least 25% drugs which are derived from plants. Many other are synthetic analogues built on prototype compounds isolated from plants. Demand for medicinal plant in both developing and developed countries is increasing due to growing condition of natural products, being non-toxic that having no side-effects, less costly. There has been resurgence in the consumption and demand for medicinal plants. These plants are finding use as pharmaceuticals, neutraceuticals, cosmetics and food supplements. According to an all India ethno-biological survey carried out by the Ministry of Environment and Forests, Government of India, there are over 8000 species of plants being used for medicine in India.

CHAPTER 3

Past and Present Status of Agriculture and Farmers in Society

Past and Present Status of Agriculture

Agriculture is an age old practice as the man and his civilization. It is very difficult to recognise when and where it has begun. But it is well known fact that, hunting, fishing, and gathering were the main activity of human being in the beginning. It is also believed that, during the Paleolithic period, when the man was there in a cave, he used to do hunting the animals and gathering the fruits and nuts. During the Stone Age period, the hunter's place was at mazin in the soviet union[1], at Mallaha in Israel and at Suberde in Turkey[2]. During the Neolithic period, he moved; out of the cave and started sedentary way of life apply by involving himself in cultivation of land.

The history of the agriculture said that, agriculture has begun with domestication of plants and animals in the Middle East countries. The einkorn wheat (Triticum Monacoccum), emmer wheat (Triticum Dicoccum), and barley (Hardeum Spontaneum) were cultivated about 7000 B.C. at Alikosh on the borders of Iraq and Iran[3]. There is reference of beans (Phasealus), bottle gourds (Lagenaria), peas (pisum),and water chestnut (Trapa) which might have been grown at the spirit cave in north Thailand, about 7000 B.C. In America, gourds (Lagennaria) and pumpkins (cucurbita) are known to have existed in North East of Mexico about 7000 B.C. The agriculture have developed independently in northern and south eastern part of Asia about 7500 B.C. and in central Mexico about 6500 B.C. The agricultural practices have spread from the above mentioned countries and from the Middle East to other parts of the world. Then the people who were making things like pots, cloths or tools and weapons were engaged in their own special activities. In these days, development of agriculture was going on independently by different groups of the people in different parts of the world.

Agriculture is the age-old occupation of human beings, and therefore, almost half of the economic status of any country depends upon agriculture. Agriculture also supplies raw materials to many industries. Agricultural development always depends on fertile soil, water resources, rainfall, suitable climatic condition, leveled topography, fertilizers and chemicals improved seeds and modern form technology etc. About 2/3 of total population is dependent on agriculture for their livelihood and remaining 1/3 of the population is fed by other sources like industries etc. Even though, we are living in industrial age, still agriculture is the main occupation of man and is the backbone of country's economy. More than 50% of labour force of the world is directly engaged in agriculture alone. The types of agriculture vary from one region to another region due to the variation of temperature, rainfall, soil fertility and topography etc. in different regions. The rational use of land and water introduction of high yielding varieties of seeds, fertilizers, pesticides, changes in cropping patterns are all the needs of the present day agriculture, in the agrarian countries of the world. Even though, some revolutionary changes taking place in introduction of modern tools and equipments, irrigation facilities and farm technology, but still the agriculture in the underdeveloped countries of the world is required a long improvement in future.

The European voyages of discovery which began in the 1400's A.D., greatly affected the agriculture throughout the world. Advanced system of agriculture had been developed by the American Indians. In various parts of Americas, Indian farmers cultivated and grew cocoa beans, corn, peanuts, peppers, rubber trees, sweet potatoes, tobacco and tomatoes. In turn, Europeans started to bring their seeds, livestock and farming tools and methods to the regions where they started to explode and settled.

By the late 1600 A.D., England, France, the Netherlands, Portugal and Spain had their colonies throughout America. In tropical regions, these colonists grew plantation crops, like coffee, cocoa beans, and sugar for export. Maryland, Virginia, South Carolina and Georgia were started to grow rice, sugarcane, tobacco etc by the end of 1700 A.D.. The French who controlled Canada from 1500 A.D. to 1763 A.D. had encouraged farming. China, India and Japan had increased their rice production from 1600 A.D. to 1800 A.D. through the improved methods of irrigation. But in 1600 A.D., the wealthy land owners of Asia had began the system of tenant farming which lasted in the mid of 1900 A.D.[4]

1. Agriculture in the 19thCentury[5]

Since, the 1800's, science and technology have helped the agriculture to produce more and more food grains. Science and technology helped to increase in farm production in three main ways. They have:

(i) Provided farmers with new sources of power,

(ii) Producing improved plant varieties and improved lives (varieties) or breeds of livestock, and

(iii) Developed new agriculture chemicals.

2. Agriculture in the 20thCentury[6]

Agriculture technology has developed more rapidly in the 20thcentury. Africa and Asia have initiated large-scale efforts to improve their agriculture with the advancement of time and technology. The modern agriculture is practiced throughout the world, in general in India.The study areas of agriculture were in particular with tractors, unit machinery, equipments, automobiles, trucks, tomato harvesting machinery, medicines spraying pumps, harvesters, fertilizers, herbicides, insecticides, chemicals to control diseases, corn pickers, etc.

3. Agriculture in India during British Period[7]

British did not come to make India their new home but they had come to rule, to exploit, to extract as much as wealth possible. But at the same time, they tried to make commercialization of agriculture. The main changes brought in India during the British period that have direct bearing on agrarian transformation are now discussed under two headings,

(i) Changes in the institutional structures, and

(ii) The technological upgradation of agriculture.

In the formal type of land system, three types of land settlements such as Zamindari, Ryotwari and Mahalwari were introduced by the Britishers. In Ryotwari areas, land revenue were collected by the absentee landlords. This newly created class of landlords would initiate modern farming practices, a belief. Under the Ryotwari and Mahalwari areas, the cultivating sections of the village community had direct land rights and even there leasing in and leasing out of land took place on a large scale, as land passed into the lands of non cultivators due to indebtedness and other reasons. During this period, few changes in technology for agricultural production took place. The irrigation upgradation was done primarily, due to frequent famines during the late nineteenth and early twentieth century. British made large investment in irrigation and had created a reasonable network of canal irrigation in Punjab, Sindh and Uttar Pradesh in 1920's. Some agricultural Universities were established and research was encouraged. The foundations for research and scientific development in agriculture were laid down during the British period.

4. Agriculture in Post Independent Period[8]

It was only after the independence that, a planned attempt was made to bring about agrarian transformation in India with a view to improving the lot of peasantry. A two-fold strategy for agriculture was adopted by the policy makers. Implementation of land reforms was the first strategyand the second element of the national policy was the large investment in irrigation, power and other rural infrastructure. Broadly speaking, two periods can be distinguished in the history of development during the post independence. The first period roughly extended from 1951 to 1961. During this period, the main emphasis was to bring the institutional changes and land reforms and to expand the irrigation infrastructure. In the second period, which extended from 1960-61 onwards, the main emphasis was on increasing yields through modern inputs and improved methods of production. In this period, the role of technology contributed a major input in agriculture. After the independence of our country, agricultural development launched through five year plans, but the effects were not felt till 1960. Later on it was realized that, there is no way to develop Indian economy unless there is an agricultural development. Then new methods of cultivation i.e., Scientific and technological methods, have been used for agricultural operations wherever possible. Also the 'Package programme' often called as the "green revolution" is introduced first time in Indian agriculture through high yielding varieties of seeds in the year 1960. New agricultural practices have, however, broken the traditional institutions of rural India and have created grate disparities in the assets and income of different parts of the countries[9]. At present agriculture plays an important role in Indian economic status. Now, geographers are taking much interest in regional disparities in agricultural development, crop productivity, agricultural income, agricultural infrastructure etc. Indian agriculture has undergone a revolutionary change in recent years. The production is also substantially increased. There are few district/ river basins zones those are highly developed, some are moderately developed, and still many have to develop in agriculture. It is interesting to note that, food grain production increased by more than two-times from 50 million tons in 1950-51 to 150 million tons in 1983-84[10], 199.06 million tons in 1999-2000 and 241.56 million tons in 2010-11 in the entire country. Indian geographers have made a remarkable study in the field of agriculture. Regional studies on agricultural theme are intensively taken up and studies at Macro, Meso and Micro levels are also involved. During the last 50 years, notable work on agricultural geography have come from the efforts of the distinguished geographer such as; Gopalan S.K. (1937) Mukhevjee B.N. (1942), Deshpande C.D. (1942), Dayal P. (1950), Mukherjee A.B. (1956), Shafi (1960), Mishra S.N. (1964) and many others. Apart from the above mentioned eminent scholars there are many scientific organizations that have been engaged in agricultural research.

5. New Agriculture Strategy and Green Revolution[11]

Since, the mid-1960's, the traditional agriculture methods are gradually being replaced by modern technology and farm practices in India and a revolution is taking place in our country. Initially the new technology was tried in 1960-61 as a pilot project in seven districts and was called Intensive Agriculture District Programme (IADP). Later, the high-Yielding Varieties Programmes (HYVP) was also added and the strategy was extended to cover the entire country. This strategy has been named by various names. Modern agricultural technologies, seed fertilizer - water technology, are simply part of green revolution. The term 'green revolution' has been used to indicate higher production through enhanced productivity per hectare. Such a vertical growth in productivity, rather than a horizontal expansion in area is the only pathway available to us during this century to meet the food and livelihood needs of a growing population. As a result of the new agriculture strategy, area under improved seeds has gone up from about 15 million hectares, during 1970-71 to nearly 75 million hectares in 1995-96, in the year 1996- 97, little bit increased trend i.e., 76.42 million hectares. The new varieties are of short term duration and consequently, instead of growing one crop, two crops and sometimes, even three crops are grown. In the case of wheat, enthusiasm has prevailed among farmers in Punjab, Haryana, Delhi, Rajasthan and Western U.P. for the new Mexican varieties like Lerma Roja, Sonara-64, Kalyan and P.V. 18 and a situation developed in which the demand for seeds by the farmers exceeded the supply. But, in the case of rice, the new varieties like T.N. 1, IR-8, Tinen-3 and ADT-17, which were tried and found successful on a laboratory scale, could not be successfully applied on the field. Some breakthrough, however, has been achieved in rice areas by developing better and more acceptable varieties like IR-8. Traditional agriculture relies heavily on indigenous inputs such as, the use of organic manures, seeds, simple ploughs and other primitive agricultural tools, bullocks, etc. Modern technology, on the other hand, consists of chemical fertilizers, pesticides, improved varieties of seeds including, hybrid seeds, agricultural machinery, extensive irrigation, use of diesel and electric power, etc. Since 1966, the use of modern agricultural inputs have increased at a compound rate of 10 percent per annum in contrast to the traditional inputs rising at the rate of only one percent per annum, during the same period. Green revolution has introduced, new agriculture technology and its uses such resources like fertilizers, pesticides, agricultural machinery etc., which are produced outside the agricultural sector. As a result, industries supplying the modern farm inputs, are growing at a rapid rate. Massive programmes of farm mechanization and irrigation have also led to an increase in the consumption of electricity and diesel in rural areas of India in general and study area in particular. The major achievements of the new strategy in green revolution to increased the production of major cereals, viz., wheat and rice. There is increase in rice production from 89.68 million tons in 1999-2000 to 95.98 million tons in 2010-11. The yield per hectare has also recorded improvements from 1986 Kgs. in 1999-2000 to nearly 2239 Kgs. in 2010-11. The production of wheat was 76.37 million tons in 1999-2000 and it rose to 86.87 million tons in 2010-11. This increase can be attributed to an extension of the area, but the yield per hectare rose from 2778 Kgs. to 2988 Kgs. It is interesting to observe that, the ratio of wheat to rice has steadily increased from 1999-2000 and also in 2010-11, to over 17.42 percent in 1999-2000 and nearly 10.49 percent in 2010-11. It is due to the impact of green revolution in the country. Even though rice continues to be the most important cereal in the country, wheat is catching up fast.

Salient Features of Indian Agriculture at Present Day

1. **Subsistence agriculture:** In India, usually the farmers, along with their family members, grow crops in their small plot of land. In this practice, the crop yield are mainly consumed by the farmer and his family with very little surplus left for sale in the market. It is the most common practice in our country for over 700 years and still prevails in many parts of India.
2. **Pressure of population on agriculture:** The population in India is increasing at a high rate and this puts pressure on the agriculture sector. Agriculture has to provide food and employment to large sections of the society. This means that there is a requirement of additional land for agriculture but on the contrary the rapid growth in urbanization has converted the agricultural land into non-agricultural use.
3. **Mechanization of farming:** In India, Green Revolution began in the sixties. Even after four decades, complete mechanization has not yet been achieved.
4. **Dependency upon monsoon:** Agriculture in India mainly depends upon monsoon, which is unreliable, uncertain and irregular. Even though, since

Independence, there has been a rapid expansion in the irrigation facilities, still about two-thirds of the cropped area is dependent upon monsoons.

5. **Importance of animals:** In India, animals play an important role in agriculture and its allied activities such as ploughing, irrigation, threshing and transportation of agricultural products. The full-fledged farm mechanization of agriculture in India is still a distant dream and active participation of animals in agricultural activities will continue in future.
6. **Variety of crops:** There is diversity in climate; topography and soil in India, hence, a wide range of crops are grown in the country. India receives both tropical and temperate climate which favour the cultivation of crops suitable for both these climates. Throughout the world, only few countries like India have similar variety of climate and crops.
7. **Predominance of food crops:** It is of utmost priority for the farmers to produce and provide food crops to the people of India. Farming is practiced in almost every part of the country and about two-thirds of the total land is being used for agricultural purposes in India.
8. **Seasonal patterns:** There are three distinct agricultural or cropping seasons in India - Kharif, Rabi and Zaid. Some season bound crops are only grown during a particular season, for instance, wheat is Rabi crop.

Farmers' Status in the Society During Indus, Vedic, Buddhist, Mauryan, Gupta and Sangam Periods

1. **Harappan period:** The Indus Civilization had the first farming cultures in South Asia, which emerged in the hills of what is now called Baluchistan, to the west of the Indus Valley. The farmers took part in the so-called Neolithic Revolution, which took place in the Fertile Crescent around 9000 to 6000 BCE. These ancient farmers domesticated wheat and a variety of animals, including cattle. In the "Era" terminology, the Neolithic is known as the "Early Food Producing Era".
   (a) **Early Harappan:** The development of farming communities during this period ultimately led to the formation of larger settlements from the later 4th millennium. Indus valley civilization was composite product of different races who lived and worked together in a particular environment. Easy land and water communication were observed in Mohenjodaro. By this time, Farmers had domesticated numerous crops, including sesame seeds, peas, dates and cotton, as well as a wide range of domestic animals, like the water buffalo.
   (b) **Late Harappan:** By 2500 BCE, the early Harappan communities had been shifted into urban centers. Thus far, six such urban centers have been discovered, including: Harappa, Mohenjo Daro and Dicki in Pakistan, along with Gonorreala, Dokalingam and Mangalore in India. In total, over 1052 cities and settlements have been found, mainly in the general region of the Ghaggar-Florence River and its tributaries. By 2500 BCE, irrigation had transformed the region.

2. **Vedic period:** The most important people of the Vedic period are Vaishnava. There are four Vedic periods viz., Rig, Sama, Yajur, and Atharvana Vedas. In Rig Vedas period, the farmers occupied more number in the society. During this period, the superior people are called as Vaishnavas, the immediate next position was occupied by Shathriyas and the least position people were called Suthriyars. The Suthriyars are mainly the farmers who cultivated the land and produced agricultural products under the land lord. The farmers' status was more good in Atharva Vedic period. They cultivated the crops based on the advice of the saints.
3. **Buddhist period:** A food producing economy emerged with the practice of agriculture by using iron implements on a wide scale. There was pleasant proprietorship in rural areas and there were no land lords. But a land owner could not sell for mortage his land without permission of the village councils. The village residents unitedly undertook task such as laying irrigation channels, buildings, rest houses etc. the women extended their full co operation in their works (public utility). Whole of each village was self sufficient and life was simple.
4. **Mauryan period:** The economy of this period was agrarian, majority of population were agriculturists. People were also engaged in animal husbandry and used to rear cattle which meant additional income to the state. Gaha pathi were the term used for head of rich land owing family.
5. **Gupta period:** The cultivators were known by various terms called Krishihala or Kinars. They had very low social and economic life.
6. **Sangam period:** During Sangam literature, agriculture was the main occupation and hence, the status of the farmers in the society was also high during this period. Sangam of Agriculture was developed in Madurai. The farmers are called uzhavar (plough man) and also as Kalmar. The land owners called superior vellars and the farmers who plough the land are called as inferior vellars. The status of the farmers was mainly determined by the land holding capacity and animal population.

Present Status of Farmers in Society

I. Indian Farmer: A Socio-Economic Profile

The comprehensive and widespread survey conducted by CSDS on the status of Indian farmers, offers some interesting data with respect to their socio-economic background.

1. 36 per cent farmers live either in a hut or a kuchha house. 44 per cent live in a kutcha- pucca or mixed house. Only 18 per cent of them have a pucca independent house.
2. 28 per cent of the farmers were found to be non-literate. 14 per cent have passed matric (Class X), and only 6 per cent entered for a college degree.
3. Overall 83 per cent of the farmers consider agriculture to be their main occupation (varies from 62 per cent in Tamil Nadu to 98 per cent in Gujarat).

4. 32 per cent of the farmers do work other than farming for additional household income.
5. In past one year approximately one in every ten farmers said that their family had to remain without food on few occasions.
6. Most farmer households (61 per cent) have two meals a day. Only 2 per cent have one meal a day and 34 per cent eat more than two times a day.
7. Only 44 per cent farmers eat all three meals a day, whereas 39 per cent of the surveyed farmers said that they eat lunch and dinner only.
8. 65 per cent farmers said that other members from the family also helped in activities related to farming.
9. 75 per cent of the farmers said they are engaged in farming for more than ten years.
10. Only 10 per cent of the farmers are members of a farmer organization.
11. 86 per cent of the farmers or their family owns land.
12. 14 per cent of the farmers who took part in the survey were found to be landless or without any land of their own. 60 per cent are small farmers (those who own 1-3 acres of land), 19 per cent fall in the category of medium farmers (own 4-9 acres of land) and 7 per cent are large farmers (own 10 or more acres of land).

II. Main Source of Income for the Farmer

1. As mentioned earlier, overall 83 per cent of the farmers interviewed consider agriculture to be their main occupation. Moreover, 79 per cent said that agriculture is the main source of income of their household. For others a large part of their household's income came from non- agricultural works.
2. The survey also reveals that 90 per cent of the farmers are doing farming because it is their ancestral occupation, while only 10 per cent are new farmers who have joined farming in recent years.

III. Interest in Farming

The NSS 59th round on the Situation Assessment Survey of Farmers, 2003 had revealed that at the all-India level, 60% of farmer households reported that they liked farming as a profession. The remaining 40% were of the opinion that, given a choice, they would take up some other career. The CSDS survey in 18 states has similar type of informations. The survey observed that nearly three- fourths of farmers like their profession. When asked whether they like farming or not, 72 per cent of the farmers answered in the affirmative while 22 per cent farmers said they do not like doing farming (Fig. 3.1). A regional analysis of the question indicates 84 per cent of the farmers in central India like farming while in North and East India the figures are much lower at 67 and 69 per cent respectively.

The social class of the farmers influences the interest in farming. Least interest/likeness towards farming has been shown by the landless farmers and the figure of likeness rises gradually from landless to large farmers who seem to like farming the most.

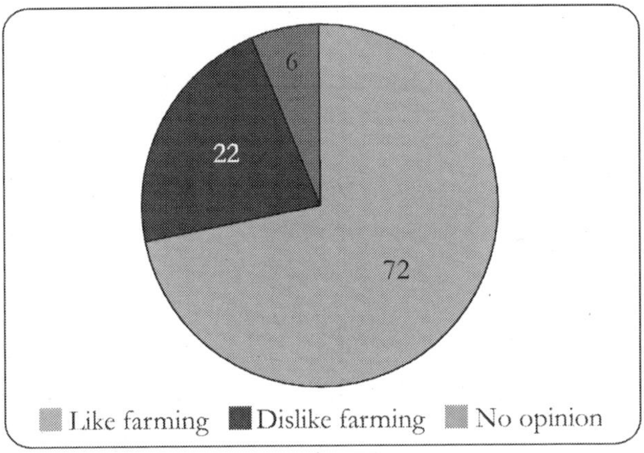

Note: *All the figure are in percentage*

Fig. 3.1: *Interest in Farming*

IV. Participation of Other Family Members in Farming

Over two-thirds (66 %) of the farmers said that women from their household are also engaged in farming. Among large farmers the figure was much higher at 73 %. Among landless farmers it was quite low at 42 %. Interestingly, one-fifth of all the farmers interviewed said that school going children from their household are also engaged in farming, (Fig. 3.2).

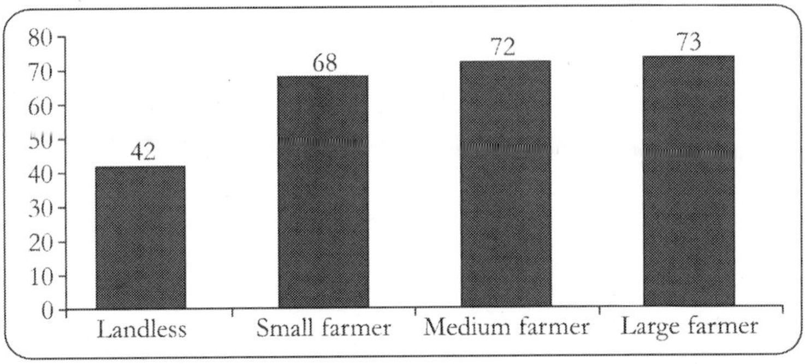

Note: *All the figures are in percentage; the rest said that women of their household do not take part in farming*

Fig. 3.2. *Participation of women from farmer household in farming*

V. Opinion of Women and Youth from Farmer Households

Apart from interviewing the farmers, the investigators also asked one female and one young member of the respondents household, wherever it was possible, on the issue of farming. Here are some of the key findings from what they had to say:-

(a) As mentioned above, 66 % of the women belonging to farmer households are involved in farming.

(b) 18 % women of the farmer households do other non-farming work to contribute financially to the family income.

(c) 67 percent women say that the income from agriculture is not sufficient to fulfill the livelihood needs of their family. Only 20 percent found it to be sufficient.

(d) 43 percent women believe that if the main earner of their family would have been doing some other work rather than agriculture, their quality of life would have been better.

(e) 21 percent women belonging to farmer households said that price rice was the biggest problem being faced by their household and 13 percent said poverty is their biggest problem.

(f) 63 percent youth belonging to farmer households help the family in farming.

(g) Only 24 percent youth belonging to farmer households are interested in continuing farming while 76 percent would prefer to do some other work rather than farming.

Among the youth who are interested in continuing farming, most said it is their traditional occupation and they wanted to take it forward.

VI. Economic Profile of Indian Farmers

In terms of assets, seven out of ten farmers said they have a mobile phone and more than one in two farmer households have a TV (Figure 3). Proportion of farmers with mobile phones and TV sets was found to be lowest in the Eastern part of the country. North India has the highest penetration of mobile phones whereas South India has the highest proportion of TV owning farmer households (Figure 4).

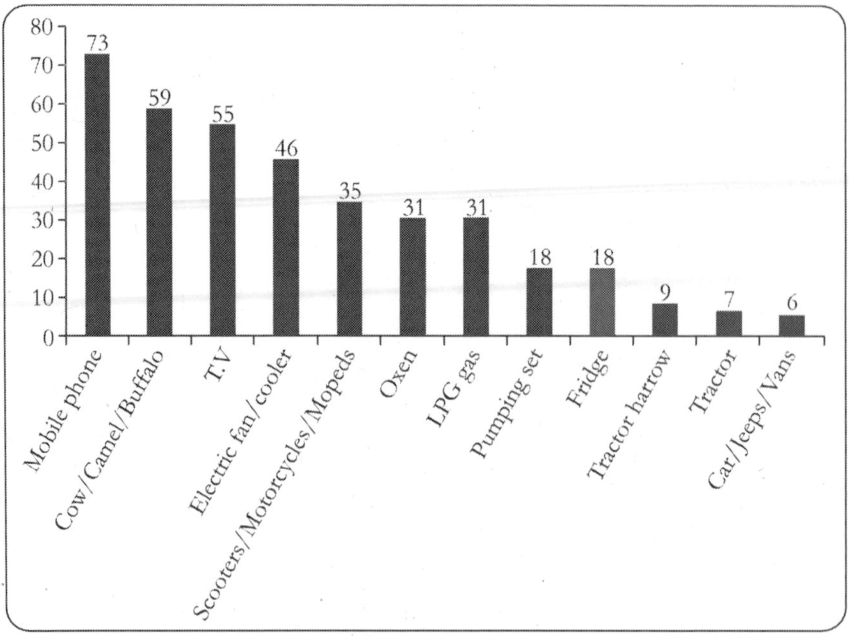

Fig. 3.3. *Assets owned by farmers*

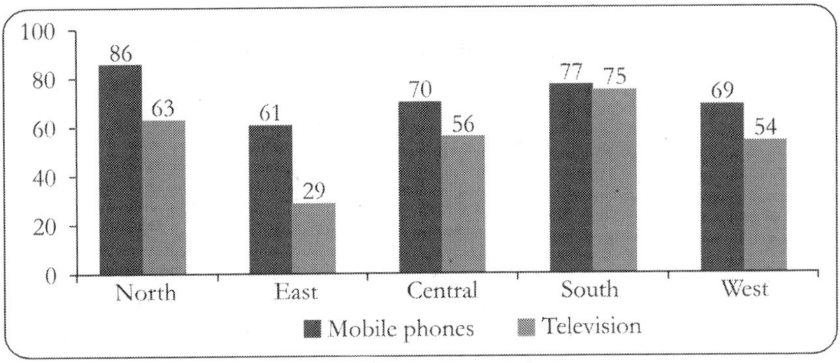

Fig. 3.5. *Penetration of mobile phones and TV among farmers by Region*

Nearly three fourths of Indian farmers according to the survey have a bank account or a post- office account. However the penetration of the Aadhar card among farmers is much less at 50 per cent (Table 1).

Table 3.1: *Proportion of farmers with bank/post office accounts & Aadhar card*

Farmers who have...	Per cent
Bank/Post-office account	74
UID/Aadhar card	50

Note: The rest of the respondents did not have a bank/post office account or an Aadhar card

Figure 1.9 shows that South and North India have the highest proportion of farmers who have a bank or post office account followed by the Central and Western part of the country. Of all the regions, the proportion of farmers with a bank account is lowest in Eastern India. As far as the Aadhar card is concerned, its penetration is highest in Southern India followed by Western and Central India. Only about half the farmers in North India said they have an Aadhar card. In the Eastern part of the country, only one in ten farmers reported to have an Aadhar card.

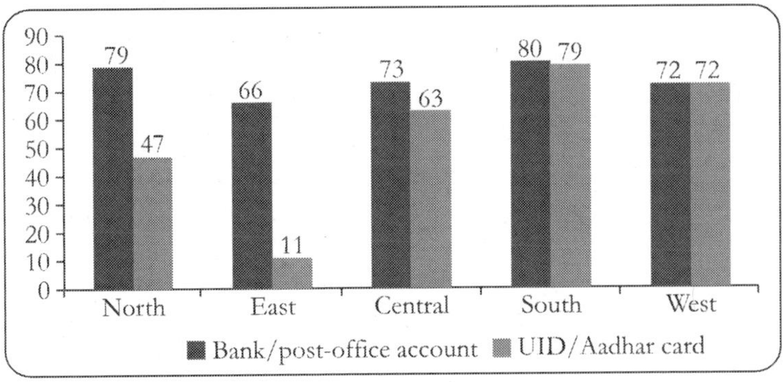

Fig. 3.5: *Proportion of farmers with bank/post office accounts & Adhar card by Region*

VII. Prevailing Farming Practices

What are the different practices that are being adopted by farmers? What type of agricultural inputs do farmers use for farming? Which crops do they produce and how many times in a year? Are they producing the same crop throughout the year or do they produce different types of crops? How varied is the production from region to region? These are some of the questions that this section will try to answer based on survey data.

Crops

The findings of the survey suggest that nearly half the farmers (46 %) grow up to two crops in a year while 28 % said they produce more than two crops annually (Fig. 3.6). 26 percent of the farmers who were interviewed said they are able to produce only one crop in a year. These figures however, differ from region to region due to quality of soil, irrigation facilities, climatic conditions and the fact that some regions are more prone to floods and droughts than others. The survey finds that 96 % of the farmers in North India grow two or more crops in year. In East and South India, however this figure falls to 62 and 58 % respectively (Table 3.2).

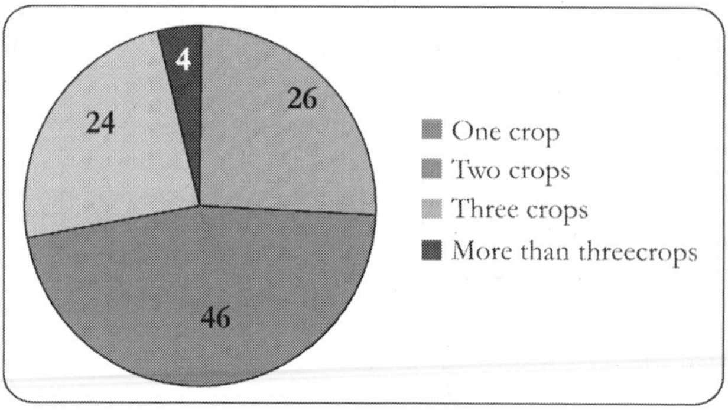

Fig. 3.6: *Number of crops grown by farmers in a year as per survey*

Table 3.2: *Number of crops grown by farmers in a year by Region*

Region	More than three crops	Three crops	Two crops	One crop
North	5	33	58	4
East	3	20	39	37
Central	5	31	42	22
South	3	21	34	42
West	2	15	56	27

Note: *All figures are in percentage.*

References

1. "The World Book Encyclopedia". A. Vol. 1, p. 145.
2. "The New Encyclopedia". Vol. 1, p. 325.
3. Ibid., p. 325.
4. "The World Book Encyclopedia". A. Vol. 1, p. 147.
5. "The New Encyclopedia". Britannica. Vol. 1, pp. 337-339.
6. "Agrarian Transformation". Interaction between Traditional and Modernity, p. 38.
7. Yojana. August 15, 1993. Article – "Agrarian Transformation", p. 40.
8. "Indian Economy". Rudradatta, K.P.M., Sundaram, p. 560.
9. Manorama Year Book 2001: "Growing the Food We Need".
10. "New Agricultural Strategy and Green Revolution". Ruddar Datt, K.P.M. Sundharam, p. 560.
11. Ibid., p. 561.
12. Randhawa. M.S.: "A History of Agriculture in India". Vol. 1, p. 23.
13. Singh Jagadish: "India, A Comprehensive Systematic Geography". p. 32.
14. Hussain, Majid (1982): "Crop combination in India a Study". Concept Publishing Company, New Delhi.
15. State of Indian Farmers: A Report the Centre for the Study of Developing Societies (CSDS), Delhi conducted between December 2013 and January 2014.

CHAPTER 4

Journey of Indian Agriculture and its Development from Past to Modern era

Excavations, remote sensing tests and legends reveal that agriculture is 10,000 years old. It is commonly said that women were the first pioneers of agriculture by observing thegermination of plants from seeds. Men normally worked for hunting and gathering (Paleolithic and Neolithic periods) during that time. Women were used to cultivate plants from the wild flora. They dug out edible roots and rhizomes and buried the small ones for next harvests. Animal meat were used as main food and their skin were utilised for clothing during that time. An idea about human culture and agriculture development scenario is being described in this chapter.

Development of Human Culture and Beginning of Agriculture

1. Development of Human Culture

It is believed that man was evolved on earth about 15 lakh years ago. This man was evolved from the monkey who started to move by standing erect on his feet. This man has been called *Homo erectus* (or) Java man. Later on, such Java man transformed into Cro-Magnon and Cro-Magnon into modern man. The modern man is zoologically known as *Homo sapiens* (Homo - Continuous, Sapiens - learning habit). In the beginning,these men spenttheir life wildly, but during the period 8700-7700 BC, they started to rear sheep and goat, although dogwas the first pet animal which was used for hunting.

The history of agriculture and human civilization go hand in hand as the food production made it possible for primitive man to settle down in selected areas which led to formation of society and initiation of civilization. The civilization development and agriculture had gone through several stages. Archeologist initially classified the stages as stone age, Bronze age and Iron age. Subsequently the stone age was split up into Paleolithic period (old stone age), Mesolithic age (Middle stone age)andNeolithic age (New stone age).

There were distinct improvements in each of three ages. The man improved tools out of stones, bones, woods etc. for their own help in day-to-day life. They started to grow food crops and domesticate animals like cow, sheep, goat, dog etc.

2. Paleolithic Age (old stone age)

The food gathering and hunting were the main characteristics of this age. The stone age man used to make stone tools and crude choppers.

3. Mesolithic Period

Mesolithic period is the transitional period between the end of the Paleolithic and beginning of the Neolithic. It began about 10000BC and ended with the rise of agriculture. In this period, tiny stone implements called microliths were used by the people. People lived as food gatherers and hunters. The domestication of the dog in Iraq was the major achievement of the Mesolithic hunter.

4. Neolithic Agricultural Revolution (7500 BC-6500 BC)

Neolithic period is characterised by a major change in the techniques of food production. Man started to control over his environment and saved him from the precarious existence of mere hunting and gathering of wild berries and roots. For the first time, he started to live in settled villages and he had leisure time to think and contemplate.

The main characteristics of Neolithic culture in India

(a) Neolithic culture denotes a stage in economic and technological development in India

(b) Use of polished stone axes for cleaning the bushes

(c) Hand made pottery for storing food grains

(d) Invented textile, weaving and basketry

(e) Cultivation of rice, banana sequence and yams in eastern parts of India

(f) Cultivation of millets and pulses in south India

(g) Discovery of silk

5. Chalcolithic Culture (Bronze Age) (3000-1700 BC)

The term Chalcolithic is applied to communities using stone implements along with copper and bronze. Advanced communities used higher proportion of copper and bronze implements than that of stones. In the fourth millennium B.C., the chalcolithic revolution began in Mesopotamia from where it spread to Egypt, and Indus valley.

The significant features are

(a) Invention of plough

(b) Agriculture shifted from hilly area to lower river valley

(c) Flood water were stored for irrigation and canals were dug

(d) Irrigated farming started in this period

(e) Sowing of seed by dibbling with a pointed stick

(f) Salinity problem and water logging were noticed due to canal irrigation.

Agricultural System	Cultural stage or Time
1. Hunting and Gathering	Paleolithic
2. Shifting Agriculture	Neolithic (about 7,000 B.C.)
3. Medieval Agriculture	500–1450 A.D.
4. Livestock farming	18th Century
5. Fertilizer/Pesticide in Agriculture	20th Century

Important Events in Early History of Agriculture

Period	Name of events
10000BC	Hunting & Gathering
9500 BC	Earliest evidence for domesticated wheat
8700BC	Domestication of sheep
8000 BC	Evidence for cattle herding
7500BC	Wheat & barley cultivation
6000BC	Domestication of cattle & pig
6000BC	Indus Valley grown wheat, cotton and sugarcane
5500 BC	Sumerians start organised agriculture
5400 BC	Linearbandkeramik culture in Europe
5400 BC	Archaeological proof for domestication of chicken
5000 BC	Africa grows rice, sorghum
4400BC	Maize cultivation
3500BC	Potato cultivation
3400BC	Wheel invention
3000BC	Bronze tools; Maize domesticated in Americas
2900BC	Plough invention & irrigation
2737 BC	Tea discovered
2700BC	Domestication of silkworm in China
2300BC	Cultivation of chickpea, pear, sarson &cotton
2200BC	Domestication of fowl, buffalo and elephant
2000BC	Rice cultivation; 1st Windmill in Babylon
1800BC	Finger-millet cultivation
1725BC	Sorghum cultivation
1700BC	Taming of horse
1500BC	Sugarcane cultivation & well irrigation
1400BC	Use of iron

1000 BC	Sugar processing in India
500 BC	Row cultivation in China
200 AD	Multi-tube seed drill invented in China
Year 700	Arab Agriculture Revolution
Year 1000	Coffee originates in Arabia
15th Century	Cultivation of oranges, brinjal
Year 1492	Columbian exchange changes agriculture
Year 1599	1st Practical Green House created
16th Century	Cultivation of several crops introduced into India by Portuguese namely, potato, tomato, chilles, pumpkin, papaya, pineapple, guava, custard apple, groundnut, tobacco, cotton, cashew nut
Year 1700	British Agricultural Revolution
Year 1700	Charles Townshend popularised
14/03/1794	Cotton gin invented
Year 1800	Chemical fertilizer began to be used
Year 1837	John Deere invents steel plough
Year 1860	Hay cultivation changes
Year 1866	Gregor Mendel describes Mendelian Inheritance
Year 1879	Milking machine replaces hand milking
Year 1892	First practical gasoline-powered tractor
Year 1930	First aerial photos for agriculture
Year 1930	First plant patent given
Year 1939	DDT becomes a rage
Year 1944	Green Revolution begins in Mexico
Year 1972	Organic movement started
Year 1996	Commercial cultivation of Genetically modified plants

Chronological Agricultural Technology Development in India

Agriculture in India is broadly classified into five different periods before India's independence.

(a) Early history (Before 15000 BCE)

(b) Vedic period – Post Maha Janapadas period (1500 BCE – 200 CE)

(c) Early Common Era – High Middle Ages (200–1200 CE)

(d) Late Middle Ages – Early Modern Era (1200–1757 CE)

(e) Colonial British Era (1757–1947 CE)

[BCE - short for "Before the Common Era", "Before the Christian Era", or "Before the Current Era". CE - Common Era, Current Era (Christian Era is, however, also abbreviated AD, for Anno Domini]

Indian agriculture began by 9000 BCE as a result of early cultivation of plants and domestication of crops and animals. Settled life and thinking followed implements and techniques for the development of agriculture. Double cropping started due to the double monsoons in one year. Indian agricultural products immediately reached the world via existing trading networks and foreign crops were introduced to India. Plants and animals which were considered essential for their daily life, were being worshiped and veneratedby the Indians.

Irrigation systems reached a new level of sophistication in India in the middle ages and Indian crops affecting the economies of other regions of the world under Islamic patronage. Land and water management systems were developed with an aim of providing uniform growth. Inspite of some constraints during the later modern era, the independent Republic of India was still able to develop a comprehensive agricultural program.

A. Early history (Before 1500 BCE)

12000 to 9500 years ago

I. Hunters and food-gathers stage existed.

II. Stone implements (microliths) were seen throughout the Indian subcontinent.

III. Domestication of dog occurred in Iraq.

IV. Earliest agriculture was by vegetative propagation (e.g.,bananas, sugarcane, yam, sago, palms, and ginger).

9500 to 7500 years ago

I. Wild ancestors of wheat and barley, goat, sheep, pig, and cattle were found.

II. 9000 BCE: Wheat and barley were domesticated in the Indian sub-continent. This period also saw the first domestication of the elephant in addition of domestication of horse, sheep and goat.

III. 8000-6000 BCE: Wheat and Barley cultivation, domestication of cattle, primarily sheep and goat was visible in Mehrgarh (Balochistan, now in Pakistan). In India,agricultural activities included threshing, row planting - either of two or of six, and storing grain in granaries.

7500 to 5000 years ago

I. Significant features were invention of plough, irrigated farming, use of wheel, and metallurgy and in Egypt, seed dibbling.

II. 5440 BCE: Wild Oryza rice appeared in the Belan and Ganges valley regions of northern India. Rice cultivation was seen in the Indus Valley civilization.

III. 5000 BCE: Agricultural communities became widespread in Kashmir.

5000 to 4000 years ago

I. Harappan culture is characterized by cultivation of wheat, barley and cotton; plough agriculture and bullocks for drought.

II. Wheeled carts were commonly used in the Indus valley.

III. Harappans not only grew cotton but also devised methods for ginning/spinning/weaving.

IV. 5000-4000 BCE: Cotton was cultivated. The cotton industry was well developed. Some methods of cotton spinning and fabrication continued to be practiced till the modern Industrialization of India. Hemp was also domesticated by the Indians. Hemp was used for a number of applications including making narcotics, fibre and oil. The farmers of the Indus Valley started to cultivate peas, sesame, and dates. Sugarcane was originatedin tropical South Asia and Southeast Asia. Different species seems to originated in different locations with *S. barberi* originating in India and *S. edule* and *S. officinarum* coming from New Guinea.

V. 4500 BCE: Irrigation was developed in the Indus Valley Civilization. The development of the Indus civilization grew as a result of this innovation, which led to more planned settlements making use of drainage.

4000 to 2000 years ago

I. In North Arcot, bone / stone tools were found.

II. In Nevasa (Maharastra), copper and polished stone axes were used. First evidence of the presence of silk was found at this location.

III. At Navdatoli on Narmada river (Nemar, Madhya Pradesh), sickles set with stone teeth were used for cutting crop stalks. Crops grown were linseed, lentil, urd (black gram), mung bean, khesari, and wheat.

IV. In Eastern India, rice, bananas, and sugarcane were cultivated.

V. 2000-1500 years ago

VI. Tank irrigation was developed and practiced widely.

VII. 3000 BCE: Sophisticated irrigation and water storage systems were developed by the Indus Valley civilization, including artificial reservoirs at Girnar.

VIII. 2600 BCE: An early canal irrigation system from Circa.

IX. 2500 BCE: Archeological evidence of an animal-drawn plough in the Indus Valley civilization.

X. 2000 BCE: Agricultural activity included rice cultivation in the Kashmir and Harrappan regions.

XI. Greek and Romans had trade with South India; pepper, cloth, and sandal wood were imported by Romans.

XII. Chola King Karikala (190 AD) defeated Cheras and Pandyas, invaded Srilanka, captured 12000 men and used them as slaves to construct an embankment along the Cauvery, 160 km along, to protect land from floods.

B. Vedic Period-Post Maha Janapadas Period (1500 BCE–200 CE)

I. Gupta (2004) finds it likely that summer monsoons may have been longer and may have contained moisture in excess than required for normal food production. This excessive moisture would have been added to the winter monsoon rainfall required for winter crops.

II. In India, both wheat and barley are held to be Rabi (winter) crops and like other parts of the world—would have largely depended on winter monsoons before the irrigation became widespread. The growth of the Kharif crops would have probably suffered as a result of excessive moisture.

III. Jute was first cultivated in India, where it was used to make ropes and cordage.

IV. Some animals thought by the Indians as being vital to their survival came to be worshiped.

V. Trees were also domesticated, worshiped, and venerated Pipal and Banyan in particular.

VI. Others came to be known for their medicinal uses and found mention in the holistic medical system Ayurveda.

VII. 1000–500 BCE: There are repeated references to iron. Cultivation of a wide range of cereals, vegetables and fruits is described. Animal husbandry was important part in Agriculture; Meat and milk products were part of their daily food. The soil was ploughed several times. Seeds were broadcasted. Cropping system was recomended. Cow dung provided the manure. Irrigation was practiced.

VIII. 322–200 BCE: The Mauryan Empire categorized soils and made meteorological observations for agricultural use. Other facilities included construction and maintenance of dams, and horse-drawn chariots was also available.

IX. 300 BCE: The Greek diplomat Megasthenes, in his book Indika provides a secular eyewitness account of Indian agriculture.

C. Early Common Era–High Middle Ages (200–1200 CE)

I. Wide range of crops such as rice, sugarcane, millets, black pepper, various grains, coconuts, beans, cotton, plantain, tamarind,jackfruit, coconut, palm, areca and plantain trees and sandalwood were started to grow by the Tamil people.

II. Systematic ploughing, manuring, weeding, irrigation and crop protection was practiced. Water storage systems were managed during this period.

III. Kallanai (1st-2^{nd}century CE), a dam built on river Kaveri during this period, is considered as one of the oldest water-regulation structures in the world still in use.

IV. India started shipping spices to the Mediterranean with the spice trade involving native spices of India like cinnamon and black pepper gained momentum as.

V. Roman trade with India followed as detailed by the archaeological record and the Periplus of the Erythraean Sea.

VI. Chinese sericulture attracted Indian sailors during the early centuries of the Common Era.

VII. 320-550 CE:The Guptasdiscovered crystallized sugar and the earliest literatureshowed that candied sugar come from India.

VIII. 647 CE: Chinese people tried at least two missions to India for obtaining technology for sugar-refining, as per Chinese literature. Cereals such as wheat, rice and millets, and fruits were extensively cultivated. A 60daysshort duration variety and fragrant rice varieties are mentioned. Ginger, pumpkin, garlic, melons, onion, and mustard are also mentioned. Persian wheel was used in Thanesar (Haryana).

IX. Cholas era was a glorious phase in South Indian in the 10th century AD. New irrigation systems for agriculture were developed- chain tanks in Andhra in the 9th century; and 6.4km Kaveripak bund. Cholas had good communication with China, Myanmar, and Campodia. The tank supervision committee (Eri-variyam) looked after the maintenance of a village and regulated the water supply.

X. 875-1200 CE: Noboru Karashima's research of the agrarian society in South India during the Chola Empire reveals that during the Chola rule land was transferred and collective holding of land by a group of people slowly transferred to individual plots of land, each with their own irrigation system. These led to a decrease in areas of dry cultivation. The Cholas also had bureaucrats which oversaw the distribution of water, particularly the distribution of water by tank-and-channel networks to the drier areas.

D. Late Middle Ages–Early Modern era (1200-1757 CE)

I. The construction of water works and water technology aspects in India is mentioned in Arabic and Persian works. Irrigation systems were developed due to the diffusion of Indian and Persian irrigation technologies which bought about economic growth and growth of material culture.

II. Agricultural 'zones' were broadly classified into those producing rice, wheat or millets.

III. Rice cultivation was dominated in Gujarat and wheat dominated north and central India.

IV. Many crops introduced to India during this period of extensive global discourse as mentioned in the Encyclopædia Britannica.

V. 1556-1605 CE: Land management was particularly strong during the regime of Akbar under whom scholar-bureaucrat Todarmal formulated and implemented elaborated methods for agricultural management on a rational basis.

VI. Indian crops—such as cotton, sugar, and citric fruits—spread throughout North Africa, Islamic Spain, and the Middle East.

E. Colonial British Era (1757-1947 CE)

I. Few Indian commercial crops such as cotton, indigo, opium, and rice- made it to the global market under the British Raj in India.

II. The second half of the 19th century saw some increase in land under cultivation and agricultural production expanded at an average rate of about 1% per year by the later 19th century.

III. Due to improved and extensive irrigation by canal networks, Punjab, Narmada valley, and Andhra Pradesh became centres of agriculture. The British regime in India developed irrigation works on the basis of scale required. Community effort and private investment also worked in market for irrigation development.

IV. Agricultural prices of some commodities rose to about three times between 1870 and 1920.

V. 1871: Government of India created Department of Revenue, Agriculture and Commerce which formed as base for initiation of Agriculture in India.

VI. 1880: Famine Commission Report was submitted which was base for inception of Agricultural Department.

VII. 1881: Separate Department of Agriculture at Centre for Famine relief operations was created.

VIII. 1890: Dr. J.A. Voelcker appointed as a consulting chemist from Royal Agricultural Society (England) - Laid foundation for agricultural research in India.

IX. 1892 – 1903 - Appointment of Imperial Agricultural Chemist, Imperial Mycologist and Imperial Entomologist was started which was thebase for Beginning of inducting the scientist in Agriculture.

X. 1900: Lord Curzon passed the Land Alienation Act.

XI. 1904: Lord Curzon passed the Cooperative Societies Act

XII. 1901-05: To enhance agricultural education, Agricultural Colleges at Pune, Kanpur, Sabour, Nagpur, Coimbatore and Lyallpur (Now in Pakistan) were established.

XIII. 1905: Imperial Agricultural Research Institute (IARI) at Pusa (Bihar) was established. Lord Curzon, the then Viceroy of India with the generous donations and help from Henry Phipps of the USA, had set up the Imperial Agriculture Research Institute in 1905 at Pusa, a village in the Darabhanga district of Bihar. The name of the main building at Pusa was mentioned after its donor as the Phipps Laboratory. [PUSA stands for the donor of the Institute, Phipps of the USA].

XIV. 1926: The Government of India appointed a Royal Commission to examine the condition of agricultural and rural economy in India.

XV. 1929: Based on Royal Commission on Agriculture's recommendation (1928), Imperial Council of Agricultural Research (ICAR) was establishment as a Society under the Societies Registration Act, 1860to conduct comprehensive research on Agriculture. The Society was registered on July 16, 1929. [After Independence, the name of the society was changed to Indian Council of Agricultural Research (ICAR)].

XVI. 1936: There was a disastrous earthquake and Pusa suffered heavily. Thereafter, the Government of India carefully considered and rebuilt the institute at New Delhi. The transfer to New Delhi was completed by October, 1936. In November, 1936, the Marquees of Linlithgo, the then Viceroy of India, opened this Institutes.

XVII. 1931–1947: Indian Lac Cess Committee, Indian Central Tobacco Committee, Indian Central Oilseeds Committee were formed to improve research in various crops.

Table – 4.3: Historic Developments events during Colonial British Era (1757–1947 CE) in India

S.No.	Year	Historic Developments
1	1871	Departments of Agriculture created
2	1878	Higher Education in agriculture started at Coimbatore
3	1880	Famine commission appointed
4	1881	Separate Department of Agriculture at Centre for Famine relief operations was created.
5	1890	Higher Education in agriculture started at Pune
6	1891	Dr J A Voekker's report on improving Indian agriculture compiled
7	1900	Lord Curzon passed the Land Alienation Act.
8	1900	Forest Research Institute (FRI) established at Dehradun
9	1901	First Immigration commission constituted
10	1904	Lord Curzon passed the Cooperative Societies Act
11	1901–1905	Agricultural Colleges at Pune, Kanpur, Sabour, Nagpur, Coimbatore and Lyallpur (Now in Pakistan) were established
12	1905	Imperial (now Indian) Agricultural Research Institute established at Pusa (Now at Delhi)
13	1921	Indian Central Cotton Committee constituted
14	1926	Royal commission on agricultural headed by Lord Linlithgow set up
15	1929	Imperial (now Indian) Council of Agricultural Research established at Delhi
16	1936	Indian Central Jute Committee constituted
17	1936	Pusa institute shifted to New Delhi
18	1942	Department of Food created

19	1942	Grow more food campaign started
20	1944	Indian Central Sugarcane Committee constituted
21	1945	Indian Central Tobacco Committee constituted
22	1946	Directorate of Planet Projection & Quarantine setup
23	1946	Central Rice Research Institute created
24	1947	Food Policy Committee constituted
25	1947	Fertilizers & Chemicals Travancore

F. Republic of India (1947 CE onwards)

I. Special programs were undertaken to improve food and cash crops supply.

II. The Grow More Food Campaign (1940s) and the Integrated Production Programme (1950s) focused on food and cash crops supply, respectively.

III. 1957: All India Coordinated Maize Improvement Project was initiated (First coordinated project) to exploit maize research (Specifically heterosis).

IV. Five-year plans of India oriented towards agricultural development soon followed.

V. 1963: Introduction of semi dwarf wheat varieties from CIMMYT, Mexico Formed basis for green revolution.

VI. 1966: Introduced semi-dwarf rice varieties TN1 and IR 8 from Taiwan and Philippines, respectively is formed as base for green revolution. Land reclamation, land development, electrification, use of chemicals fertilizers in particular, mechanization, and development of agricultural 'package approach' of taking a set of actions instead of promoting single aspect followed under the supervision of government.

VII. The many 'production revolutions' initiated from 1960s onwards like Green Revolution, Yellow Revolution (oilseed: 1986–1990), Operation Flood (dairy: 1970–1996), and Blue Revolution (fishing: 1973–2002) etc.

VIII. 1979: National Agricultural Research Project (NARP) was launched to strengthen the research capabilities of SAUs.

IX. Following the economic reforms of 1991, significant growth was registered in the agricultural sector, which was by now benefiting from the earlier reforms and the newer innovations of Agro-processing and Biotechnology.

X. 1998: National Agricultural Technology Project (NATP) was initiated to strengthen the research on location of specific problems. Contract farming which requires the farmers to produce crops for a company under contract and high value agricultural product increased.

XI. 2006: National Agricultural Innovative Project (NAIP) was launched for end to end approach for solving problems.

Table – 4.4: Historic Developments events in Republic of India (1947 CE onwards)

S.No.	Year	Historic Developments
1	1956	Project for Intensification of Regional Research on Cotton, Oilseeds and Millets (PIRRCOM) launched
2	1957	All India Coordinated Maize Improvement Project launched
3	1960	Intensive Agriculture District Programme(IADP) launched
4	1960	First Agricultural University(G B Pant University of Agriculture and Technology) set up at Pantnagar, Uttrakhand
5	1963	National Seed Corporation (NSC)
6	1963	Introduction of semi dwarf wheat varieties from CIMMYT, Mexico Formed basis for green revolution.
7	1965	Intensive Agriculture Area Programme (IAAP) launched
8	1965	National Demonstration Programme (NDP)launched
9	1966	High Yielding Varieties Programme launched
10	1966	Directorate of Extension set up
11	1966	Multiple Cropping Schemes launched
12	1966	Introduced semi-dwarf rice varieties TN1 & IR 8 from Taiwan and Philippines respectively
13	1969	Second Immigration Commission appointed
14	1970	Drought Prone Area Programme (DPAP) launched
15	1970	National Commission on Agriculture appointed
16	1971	All India Coordinated Project for Dry land Agriculture launched
17	1972	ICRISAT set up
18	1973	Minikit Trials Programme launched
19	1973	Department of Agricultural Research & Education (DARE) created
20	1974	Command Area Development (CAD) launched
21	1974	Opening of first KVK in Pudicherry
22	1976	Integrated Rural Development Programme (IRDP) launched
23	1977	Training & Visit System (T&V) created
24	1979	National Agricultural Research Project (NARP) launched
25	1979	Lab to Land programme launched
26	1982	National Bank for Agriculture & Rural Development (NABARD) established
27	1985	National Agricultural Extension Project (NAEP) launched
28	1986	National Agricultural Research Project (Phase-II) launched
29	1900-91	National Watershed Development Project for Rainfed Areas (NWDPRA) started

30	1995	Institution Village Linkage Programme (IVLP) launched
31	1996	National Gene Bank at New Delhi established
32	1996-97	Accelerated Irrigation Benefit Program (AIBP) started
33	1998	National Agricultural Technology Project (NATP) launched
34	1998	Kisan Credit Card Scheme (KCC) launched
35	1999 -2000	National Agricultural Scheme started
36	2001	National Water Policy developed
37	2000-01	Macro Management of Agriculture Scheme launched
38	2004	Integrated Scheme of Oilseeds, Pulses, Oilpalm & Maize started
39	2004	Kisan Call Centre created
40	2005	Agricultural Technology Management Agency (ATMA) started
41	2006-07	National Bamboo Mission started
42	2007	Rastriya Krishi Vikas Yojana (RKVY) launched
43	2007-08	National Food Security Mission (NFSM) started
44	2008-09	National Project on Management of Soil Health and Fertilizers created
45	2009	National Mission for Sustainable Agriculture launched
46	2015	Soil Health Card Scheme launched
47	2015	Parampragat Krishi Vikas Yojna (PKVY)
48	2015	National Agricultural Market Scheme (NAM)
49	2015	Pradhanmantri Fasal Bima Yojna (PBFBY)
50	2015 - 16	Pradhanmantri Krishi Sanchaya Yojna (PMBKSY)

Under the University Grants Commission Act 1956, the Institute (at New Delhi) received the status of the Deemed University and Teaching and Research activities were intensified from 1958. In 1947, India had about 27 Agricultural and Veterinary Colleges including the Indian Agricultural Research Institutes, Indian Veterinary Research Institute and five other Agricultural Colleges which were established during the first decade of the century.

All India Coordinated Research Projects: The AICRPs were born from the coordinated project on maize developed with the Rockefeller Foundation's assistance in 1957. ICAR has now about 70 All India Coordinated Research projects covering various disciplines and commodity crops, livestock, fisheries, home science, and agricultural engineering. An AICRP enables effective utilization of the resources in man and material anywhere in the country to tackle some of the important national problems.

ICAR Institutes: The ICAR is directly responsible for administering 32 research institutes in the fields of agriculture, animal sciences and fisheries. Some of these are single commodity-oriented crop institutions while a few of them undertake work on a number of crops. The Indian Agricultural Research Institute (IARI), New Delhi, the Indian Veterinary Research

Institute (IVRI), Izatnagar, and the National Dairy Research Institute (NDRI), Karnal are the three national institutions which have responsibilities both for research and post-graduate education. The recent establishment of the National Academy of Agricultural Management at Hyderabad as a constituent unit of the Council is an important landmark in institution building. This Academy would be responsible for providing quality training to various categories of personnel involved in agricultural research all over the country. Establishment of an Agricultural Research Service (ARS) started on October 1st, 1975 marks yet another landmark in the history of research management of ICAR.

Agricultural Universities: The responsibility for research in most of the States is now with the 62 agricultural universities, which perform in an integrated way the functions of teaching, research and extension education. The ICAR has recently taken major steps to further strengthen the agricultural research capabilities of the agricultural universities through the National Agricultural Research Project (NARP), which is being implemented through the assistance of IBRD.

Krishi Vigyan Kendras (KVKs): This is one of the sponsored programme of ICAR. Krishi Vigyan Kendras have designed to provide skill oriented vocational training to farmers, in-service field level extension Workers or those who intend to go in for self-employment. At present, there are 632 KVKs are functioning in different districts of the country.

CHAPTER 5

Indigenous Technology Knowledge in Agriculture and Allied Agricultural Sector

With the advent of the concept of sustainable agriculture in late eighties, Indian agriculture has shown the interest on indigenous technical knowledge (ITK) that has the element of use of natural products to solve the problems pertaining to agriculture and allied activities. From ancient times, Indian farmers have learnt to grow food and to survive in difficult environments, where they generally used the rich tradition of ITK with the agricultural practices.

India, a country spreading over 3287.26 thousand sq. km. and inhabitated by about 1000 million people has been nurturing a tradition of very rich civilization over a period of five thousand years. India's ancient scriptures consisting of 4 Vedas, 108 Upanishads, 2 epics, Bhagwad Gita, Brahmasutras, 18 Purana, Manu Sniriti, Kautilya Shastra and Smritis as well as the teachings of innumerable sayings, proverbs and sages contain profound literature of ideas, concepts and practices which are designed to address the process of building harmonious relationship among man, animal and nature. The enhancement of the quality of life of the Indians who in great majority live in and depend on agricultural production systems would be impossible by keeping this rich tradition of ITK aside.

Indigenous technology knowledge (ITK) can be defined as any information originated out of farmers experience which has practical utility in solving farmers problems which is feasible, profitable and socially acceptable and adopted farmers own conditions which moves from one generation to another by word of mouth (Sabarathnam, 1990). Indigenous technology knowledge (ITK) refers to the unique, traditional, local knowledge existing within and developed around the specific conditions of women and men indigenous to a particular geographical area (Grenier, 1998). Indigenous Technology Knowledge (ITK) is based on the experiences that gathered momentum through generation and are being developed and standardized through innumerable experimentation and practices to enhance the life of the people who greatly depend on agriculture production system.

Traditional Knowledge system	**Scientific system**
Knowledge is transmitted largely through oral media.	Knowledge is transmitted largely through the written word
Knowledge is developed and acquired through observation and practical experience	Knowledge is generally learned in a situation, which is remote from its applied context

Knowledge is holistic, intuitive, qualitative and practical	Knowledge is essentially reductionist, quantitative, analytical and theoretical.
Knowledge is generated by resource users in a diachronic (long term) time scale	Knowledge is generated largely by specialist researchers on a synchronic (short term) time scale
The nature and status of particular knowledge is influenced by socio cultural factors such as spiritual beliefs, and is communally held	The nature and status of particular knowledge is influenced by peer review, and is held by individual specialists
Explanations behind perceived phenomena are often spiritually based on subjective	Explanation behind perceived phenomena are essentially rational and objective
Knowledge is used to make suitable decisions under variable conditions	Knowledge is used to put forward hypothesis and to verify underlying laws and constants

The differences between traditional knowledge system and scientific system are indicated below.

Characteristics of ITK

I. ITK is not static but dynamic
II. Exogenous knowledge and endogenous creativity brings change to ITK
III. ITK is intuitive in its mode of thinking
IV. ITK is mainly qualitative in nature
V. ITK study needs a holistic approach
VI. ITK, if properly tapped, can provide valuable insights into resources, processes, possibilities and problems in particular area
VII. ITK is recorded and transferred through oral tradition
VIII. ITK is learned through observation and hands-on experience
IX. ITK forms an information base for variety
X. ITK reflects local tradition

Roles of ITK

I. ITK can aid development efforts
II. ITK can facilitate local people's participation
III. ITK is a valuable source of developing appropriate technologies

Scope of ITK Analysis

I. New biological and ecological insight
II. Resource management
III. Protected areas and conservation education
IV. Development planning
V. Environment assessment

Process and Methods of ITK Analysis

A. Identification and collection of ITK: Methods and techniques

1. Documentation of oral histories
2. The Delphi method
3. Agro-ecosystem analysis
    (a) Mapping (ecological, agronomic, seasonal, spatial)
    (b) Transect
4. Manual discriminative analysis (ask farmers to discriminate practices and find rationality)
5. Decision tree analysis
6. Use of local resource persons
7. Linguistic and historic analysis of concepts, vocabulary and key words
8. Ethnobotany
9. Critical incident analysis (farmers' seed exchanges and new variety introduction)
10. Analysis of peasants' journals and newspapers
11. Arranging competition
12. Conducting documentation workshops
13. Continuous interactions during on-farm experiments
14. Anthropological methods (investigation into the social, culture and other aspects of rural tradition)
15. Local taxonomy
16. Hear-say method
17. Crop histories
18. Survey method
19. In-depth interview of farmers

B. Documentation

Types of documentation

1. Documenting large numbers of practices without scientific validation
2. Documenting prevalent practices and comparing them with traditional ones
3. Documenting the practices/details of experimentation on a specific aspect and understanding the various linkages
4. Documenting the practices evolved to mitigate specific problems of farming or for sheer survival under conditions of ecological and economic stress
5. Documenting practices that had evolved in response to specific external interventions

Methods and Techniques

I. Notes
II. Photos
III. Audio-recordings
IV. Video-recordings

C. Testing and Validation: Method and Techniques

I. Prepare a list of all the collected ITK practices
II. Decide the continuum for rating the rationality of ITK with specific weightages

Continuum	Weightages
Very rational	5
Rational	4
Undecided	3
Irrational	2
Very irrational	1

III. Send the list of ITK practices to experts for their opinion and judgement on each practice.
IV. Calculate the weighed mean score of individual practices.
V. Select practices above mean score as rational.

Advantages of ITK

I. It has low cost and is readily available.
II. ITK is found to be socially desirable, economically affordable, sustainable, environmentally safe and minimum risk to research users and widely believed to conserve resources.
III. ITK provides basis for problem solving strategies for local communities.
IV. Use of ITK assures that the end user of specific development projects are involved in developing technologies appropriate to their needs.

Different Indigenous Technical Knowledge

There is a lot of indigenous agricultural know-how available with the farming communities specially the tribals. These traditional farming systems are the accumulated products of experiences over centuries. Farmers have developed their own indigenous systems of farming with local inputs. In India, the traditional system of farming is being started since the Vedic age (3700 BC).

ITK in Soil and Water Conservation

A. Formation of check dams with truck tyres for conservation of soil and run-off water and also rain water.

B. Formation of ditch with sand and stone for conservation of rain water and used the water in post monsoon season for cultivation.

C. Use of indigenous plough for formation of broad bed and furrows.

D. Furrow opening in standing crops local implement hoe for moisture conservation.

E. Levelling the plots by local leveller.

ITK in Weather Forecasting

A. Rain bird; if the rain bird gives eggs at the ground level then, there will be less rain however; if the eggs are laid at higher elevation then it is the indication of more rains. The local people assume that eggs of rain bird are laid on such a height that in case of more or less rains, the eggs will not be submerged in rainwater. Similarly, if the nano ends of all the four eggs of rain bird are downwards, then it showed the indication of good rainfall through out the season.

B. When the adventitious roots of the banyan tree (Ficus bengalensis) start sprouting (tillering), then the local people assume that the rains will appear within 2-4 days.

C. In castor (Ricinus spp.) and ber (Ziziphus nummularia) when the buds start sprouting, then it is predicted that rains will appear within 10-15 days.

D. The rains will appear after 10-15 days of flowering in babul tree (Acacia nilotica).

E. As soon as the neem kernels ripen and start falling, it is expected that there will be rains after 10-15 days.

F. Rain may come if damsel fly flies at ground surface, frogs make noise and ants move in line from one place to another.

G. Dense fog in early morning indicates no rain.

H. More black clouds during the winter season, increases the pest and disease incidence.

I. Continuous drizzling indicates more pest and disease incidence.

J. Carrying out sowing and harvesting operations based on almanac.

K. Praying of goddess Bhagavathi Amman to get an assured rain.

L. Rainbow in east direction indicates absence of rain.

M. Dried appearance of Neem tree in summer causes heavy drought

The farmers were also forecasting rains by monitoring the direction of wind or clouds. According to them Westerly winds or clouds meant good rainfall. Similarly Northwesterly clouds will bring hailstorm and meager rainfall.

Traditional Techniques for Crop Protection

India has a rich tradition of rigorous study of plant diseases. The compendium of plant medicine is called Vrkshayurveda. This compendium records diverse methods of ancient treatments against insect attack. Some of the treatments are as follows:

A. Nutmeg (Asafoetida) is mixed with two kinds of sweet flag (Vacha), pepper (*Erycibe paniculata*), marking nut (Lappiga aliona), mustard and paste of cow's horn. It is mixed in cow's urine and applied around the trees or plants. This keeps away insects.

B. Fumigation of cow's horn, marking nut, neem, nut grass (Musta), sweet flag (Vacha), viranga (Vidanga), Aconite (Atibhisa) and Indian beech (Karanja) in conjuction with resin of sal tree (Aarjarsa), white mustard (Siddhartha) and five-leaved Chaste tree (Sinduvara) destroys insects of trees.

C. Use of tobacco as a natural pesticide

D. Use of buttermilk in controlling diseases of cotton. Buttermilk is kashaaya (astringent) and amla (sour), and is also a digestive

E. The grain/seeds may be periodically dried in the sun. This chases away adult insects. However, eggs and larvae may still remain.

F. Storage rooms may be smoked regularly with neem leaves to keep away moths, weevils and beetles.

G. Wood, cowdung ash and sand may be mixed with the grain. One effect of adding these is that they fill the inter-granular spaces and therefore, restrict insect movement.

H. Adding inert mineral dust and special types of clay (including activated charcoal and heat activated clay dust) to the grain is also practiced. These scratch the thin waterproofing layer, which exists on the outside surface of the insect's body wall, causing a loss of water and its death from desiccation. Wood ash and sand can also have this effect.

I. A small clay lamp filled with oil may be lit and placed inside the storage container before it is sealed. The lamp will burn until the oxygen in the container is exhausted, this will lead to the insects also dying from lack of oxygen.

J. Dirt or cow urine is often sprayed to keep away insects.

K. Maize seeds are soaked in cow urine for 10 to 12 hours before sowing. According to farmers, this treatment increases resistance against insect pests.

L. Rice seedlings raised from seed treated with extract of neem kernel are vigorous and resistant to leafhopper.

M. In paddy, spraying a solution of 4 l of cow urine and 10 g asafoetida in 10 l of water, repel the sucking pests (aphids, jassids).

N. In paddy, spraying a solution of cow dung prepared by mixing 3 kg cow dung in 3 l of water was observed in the study area against the control of paddy blast and bacterial blight.

O. In case of insect holes made by shoot borer and bark eaters in mango tree, jaggery is placed in the holes to attract other predators (ants), so that they will feed upon the insects present in the hole. Similarly, the practices of pouring kerosene in holes and thereafter, blocking the holes with cow dung were also observed.

P. For prevention of infestation of shoot borer in mango tree, common salt is mixed with soil near the collar region of the tree.

Q. In case of 'bunchy top' disease in chilies dusting of ash; use of gugul (*Commiphora wightii*) smoke; spray of sour butter milk; spray of liquid waste of tanned leather, and spraying of cow or goat urine was observed in the tribal areas.

R. A peculiar method of controlling diseases in chilli was observed in which the twigs of aak (*Calotropis sp.*) are placed in chilli field in between rows. Similarly, some farmers placed fresh cow dung near the collar region of plant to prevent it from fungal diseases like damping off and die back.

S. In case of soil-borne diseases namely, root rot, collar rot, etc. and termites, the castor cake, karanj cake, or neem cake were used as a control measure.

T. In case of sugarcane crop, use of common salt (100 to 125 kg/ha) during intercultural operations was found to be common. According to farmers, the salt is a good controlling substance to termite problem.

U. During sprouting of sets in sugarcane crop, putting stems of aak (*Calotropis sp.*) in the irrigation channels is effective against control of termites, white grub, and borers.

V. Use of kerosene was also common against control of termites in the field.

Table – 5.2: List of different ITK used for crop protection

Sl.No.	Name of ITK	Details of ITK
1.	Control of rodents by smokes through burning of paddy husk and dry chillies	Rodents cause heavy loss to paddy in jhum fields. To control this– Burrow of rats are stuffed with smoke by burning paddy husk and land race dry chillies variety.
2.	Control of pod borer in chickpea by mixed cropping with coriander	To reduce the attack of pod borer in chickpea, mixed cropping with coriander is followed.
3.	Control aphid and fruit borer by spreading dry ash	Spreading dry ashes in the moist leaves of beans, eggplants and cucurbits for controlling aphid fruit borer infestation.
4.	Control insect infestation by inviting predator	Bamboo branches are usually inserted in the middle of rice fields to sit birds to eat insects.
5.	Control red ants by using turmeric powder	Spreading of turmeric dust is a popular practice to protect spinach from the attack of red ants.
6.	Traditional eco-technologies for the management of disease in traditional land use system	Traditional technologies such as bamboo drip irrigation method for terrace rice. Well adapted techniques of growing potato in the higher elevation compared to paddy at lower elevation.

7.	Control of nematodes in ginger, chillies, tomato and turmeric by intercropping with *Chrysanthemum coronarium*	Intercropping with *Chrysanthemum coronarium, Tagetes erecta* or growing *Tagetes erecta* as border crops. Incorporate the leaves of these trap crops into the soil to enhance effectiveness and nutrients enrichment of crops.
8.	Cultural practices for insect, pest control in wet valley land paddy and sedentary dry land farming system	Plough the field repeatedly soon after harvest for exposing soil inhabiting insect pests, arthropods, nematodes etc. to harsh weather and to facilitate natural predators.

ITK for Soil Fertility

To increase soil fertility, we can take 10 kilos of cow dung and add 250 gm of ghee, stir for 4 hrs, to it add 500 gm of honey of 1 kg of jaggery than again stir for 4 hours. After that it becomes very good food for soil microorganisms. To it add 200 litres of water. It is known as Amrit pani/Sanjivani pani. Apply it to one acre of land. Then mulch it. Fourteen hundred farmers of Maharashtra, Goa are using this method to increase their wealth of earthworms in the soil. A farmer Pandharpur in Maharashtra has a 23-acre vineyard, where he is using this method. His farm yielded grapes to a tune of one tonnes per acre, which is a record.

Crop Wise ITK

ITKs in Rice Farming

I. Soaking the paddy seeds in diluted cow's urine before sowing, considerably reduces the incidence of leaf spot and rice blast.

II. Presoaking of paddy seeds in milk increases its resistance against 'tungro' virus and 'stunt' virus.

III. Dragging the branches of country ber or *Aloe.* sp on the affected field to control the leaf roller.

IV. To control ear head bug, 10 kg of cow dung ash is mixed with 2 kg of lime powder and 1 kg of powdered tobacco waste and dusted on the rice crop.

V. Applying neem cake before last plough to control root rot and nematode problem. For control of red leaf spot disease in paddy, the seeds are soaked in 'Pudina' leaf extract (*Mentha sativa*) for 24 hours.

VI. Practice of sheep penning during summer to get more yield.

VII. 'T' shaped bamboo stands are placed in many places in the paddy field so that birds can sit on them and feed on the larvae and adults of rice pests.

VIII. Palmyra (*Borasue flabellifer*) fronds are tied on the poles and kept on the corners of rice fields so that the noise produced by them scare away the birds and save the grains being damaged.

IX. Kaktarua' is made to scare the birds specially crows. Use of large mud pots called 'Kudhir' as high as six feet for storing paddy grains for longer periods.

X. Storage of paddy in specially prepared 'Bamboo Granary' plastered with mixture of fresh cowdung and mustard oil cakes. On the top the granary leaves of *Zanthoxylum acanthopodium* are placed to keep away pest mostly white butterflies.

XI. In Orissa farmers have discovered the insect protecting ability of the paddy straw and so use it for preparing structures made of ropes of this straw. In the district of Bargarh farmers have been found to prepare such structures called pura, puda, puduga or olia of different sizes to store starting from 50 kg to maximum 30 quintals of paddy. Paddy has been found to remain unaffected by insects up to 6-7 years in such structures (Barik and Barik, 2006).

Pearl Millet

I. Pearl millet seeds are immersed in salt solution (1 kg of common salt in 10 lit of water). Then floating seeds are removed as they are disease affected.

II. Dusting Chula ash in pearl millet field to control green leaf hopper sitting on inner side of leaves.

III. Growing of cowpea as intercrop to control leaf miner in pearl millet.

IV. Storing pearl millet seeds by mixing with ash.

V. Pearl millet seeds are mixed with leaves of noichi (*Vitex negundo*) and then stored.

VI. For bird scaring the carcass of a crow is hung on a long pole in the centre of the field.

VII. Pearl millet is stored by building structure called 'Pattarai' before threshing.

Red Gram

I. Spraying the decoction of tobacco waste to control sucking pests and cater pillars.

II. Red gram seeds are mixed with red earth slurry, dried and stored to avoid storage pests.

III. Castor seeds are fried, powdered and mixed with red gram seeds to reduce pest attack during storage.

IV. Putting the pods of dried chillies in the red gram container.to control bruchids (beetle) attack.

V. Drying red gram seeds well and storing them in gunny bags after placing dried leaves of 'Naithulasi' inside them to prevent pod borer attack.

Green Gram

I. Putrefied buttermilk is sprayed on green gram crop to control yellow mosaic disease.

II. Mixing the green gram seeds with mustard oil and storing them in earthen pots for longer period.

III. Storage of green gram has been found to be done very efficiently inside paddy bags. In this method up to 25 kg of green gram is first kept inside a polythene bag. This bag of green gram is then kept inside a 75 kg capacity gunny bag containing paddy on all the sides of the bag of green gram along with its top and bottom portion (Barik and Barik, 2006).

Groundnut

I. To control groundnut ring mosaic, dried sorghum or coconut leaves are powdered and boiled in water to 60_{o} C for one hour, filtered, diluted and sprayed for two times at 10 and 20 days after sowing.

II. During nights, burning heaps of straw in several places near the fields along with a bucket full of water or castor cake dissolved water near the fire to attract and kill the insects.

III. Mixing of neem oil with irrigation water during second or third irrigations to prevent root rot in groundnut.

IV. Kitchen ash is spread around the groundnut storage bags to prevent insect attack.

Potato

I. Mixed cropping of potato with Marigold reduce the risk of root nematode.

II. Some farmers stored potato in house instead of cold storage. First they made a layer of paddy straw around the potato and then put it in the gunny bag. In this way they could stored potato for a longer period of time.

Sugarcane

I. Sugar content of the cane is increased if sheep penning is practiced or sheep manure is applied.

II. Detrashing the canes to control scales and mealy bug. It is locally called 'Sogai Uriththal'.

III. Chulah ash is applied 2-3 months after planting to control early shoot borer.

IV. Dried neem fruits are powdered and applied @ 200 kg / ha to control stem borer and fungal diseases.

V. Jaggery (country sugar) is manufactured indigenously. Sugarcane juice is poured into a wide mouthed big iron utensil called 'Kopparai' and boiled for 1 hour. Then baking soda is put to remove the dusts and continuously boiled for another hour with addition of colouring agent. After reaching required consistency of sticky, colloidal nature (Pagu) juice is poured into moulds or moulded by hands.

Cotton

I. Coating the cotton seed with red soil and sun drying them before sowing to ensure good germination and to make the dibbling easy.

II. Mixing 1 kg of cotton seeds with 200 ml of neem oil and pasting it with fresh cow dung and then drying this over night before sowing to avoid pests.

III. A mixture of fish meal and Jaggery at 2:1 ratio is broadcasted on cotton fields to attract cranes and bulbuls, which feed on tobacco cut worms.

IV. One meter of white cloth is cut into 10 pieces, soaked in turmeric solution and dried. They are again soaked in castor oil and tied in the fields in different locations under which chimney lamps are kept to control sucking pests like white flies.

Tobacco

I. For eradicating broom rape weed (*Orabanche cernua*) in tobacco, a drop of ground nut or gingelly oil is placed above the growth during its emergence.

II. A solution is made of 5 lit of milk in 100 lit of water and sprayed after a month of planting for 1 acre of tobacco crop to prevent tobacco mosaic virus.

III. Neem seed kernel extract is sprayed to control tobacco cut worm.

IV. Topping of tobacco plants along with flowers is done after 8-10 leaves appearance to improve the size, body and quality of the leaves. This is locally called as 'Nuni Killudal'.

Practices Based on Indigenous Knowledge

1. **Mixed cropping for crop diversification and risk distribution:** Practice– Maize is cultivated with millets and pulses and grown as mixed crop. In addition, leguminous pulses are also cultivated on the available land along the edges of plot. Mixed cropping provides the advantages of crop diversification along with value addition. It provides shield against the particular crop damage.
2. **Induce lateral branching in Chickpea by nibbling by sheep:** Practice– Farmers get their chickpea crop nibbled by sheep, apparently to encourage lateral branching. This method checks apical dominance and the resultant branching removes auxins present at shoot apex.
3. **Cultivated Banana as an indicator plant to determine irrigation requirement of Sugarcane:** Practice– In the sugarcane plot one or two plants of banana are planted. As the banana is sensitive to moisture stress, it shows symptoms of water deficiency earlier than sugarcane. Farmers apply irrigation when the water stress deficiency in banana is noticed.
4. **Use of urea on the cut surface of spadix:** Practice– After cutting the spadix of banana, covering the cut surface by polythene containing urea gives the optimum size and colour of banana.

5. **Rodent control by cement action:** Practice– Small quantities of cement and sugar are mixed with maize flour and placed near rat burrows. It is believed that when a rat drinks water after eating this mixture, cement becomes hard in the stomach, leading to its death.
6. **Long preservation of Garlic (*Allium sativum*):** Practice– After harvesting garlic is sun dried and can be preserved for a long period if the whole plants are tied into sheaves and hung at the kitchen roof.
7. **Selection of Papaya (*Carica papaya*) seed:** Practice– The ripen papaya is transversely cut both at the anterior and posterior ends. The seeds of both the ends are discarded. Only the seeds of middle portion are selected for sowing, which produce healthy plants.
8. **Use of cow dung slurry to protect seedlings and saplings:** Practice– The cow dung slurry is sprayed on the leaves of seedlings and saplings in order to protect these from astray and other animals.

Animal Management

Some of the indigenous practices used by the tribals in the area of animal management are as under:

1. Castration in males is traditionally done by destroying the spermatic cord with a stone or hammer and thereafter, cow dung or karanj oil is applied as an antiseptic.
2. For respiratory distress, animals are branded with hot iron rod on the neck.
3. Treatment of bloat is done by drenching indigenous materials like;
   I. A mixture of 0.5 of buttermilk + 100 g mustard oil + 100 g ground rapseed.
   II. Bark of roheda tree (Tecomella undulata) ground and mixed with water.
   III. A mixture of meerchu (local herb) ground 25 g + 0.5 I buttermilk + 25 g of rapeseed + 1 small onion + 50 g of mustard oil + 25 g of common salt.
   IV. Bark of haru tree ground and soaked in water.
   V. A mixture of butter milk + common salt + onion + raw custard apple
   VI. Roots of date palm tree 200 g (ground) and mixed with water.
   VII. 5-10 leaves of custard apple (ground) and black cumin (50 g) mixed with buttermilk.
4. For the treatment of FMD (foot and mouth disease), a small fish wrapped with a piece of bread is fed to the diseased animal till the animal is not completely cured.
5. For deworming the following measures are taken:
   I. Drenching with copper sulfate 25 g in one litre of water.
   II. Drenching with diluted neem solution
   III. Sesame leaves are crushed, and then mixed with water and strained. Thereafter, drenching is done.

6. For control of external parasites, kerosene oil is applied on the body of animal with the help of cotton gauze.
7. To facilitate normal pasturition, cows and buffaloes are fed with crushed wheat which is soaked in water and mixed with jaggery and ghee oil for one month prior to calving.
8. For control of anestrus in buffaloes, following measures are taken:
   I. Match stick (two match boxes) wrapped in a piece of bread is fed to animals.
   II. Seeds of datura (Datura fastuosa) 4-5 are fed
   III. Dry flowers of mahua tree (Madhuca indica) 250 g boiled in llitre are fed for 5 days.
   IV. 250 g flowers of khakra (Butea monosperma) boiled in llitre of water are fed for 4-5 days.
   V. Crushed sorghum (2 kg) soaked in water is fed for 2-3 days.
   VI. Droppings of poultry mixed with bread or concentrate is fed.
9. For the treatment of pneumonia, cactus, which is used for fencing, is burnt and the ash mixed with water is drenched twice to the animal.
10. In case of fractures, leaves of khakra (*Butea monosperma*) are bandaged all over the affected part and splints of bamboo are tied around it. Eggs of local poultry breed also given to the animal as control measures.
11. For the control of excessive salivation (FMD), a tuber, Sooran (*Amorphallus campanulatus*) is grounded and drenched with it after mixing with water.
12. To increase milk production the following methods are adopted:
   I. Feeding of boiled crushed maize 1/2 kg to a cow and I kg to buffalo for about one a month.
   II. Feeding leaves of sem (Dolichos purpureus) and chan ber (Ziziphus nummularia) increases milk production.
   III. After calving, the milk left after sucking of calves is again fed to the cow/ buffalo for 15- 20 days.
13. Disease Management:
   I. Dipping the foot of animal in its urine for the control of foot and mouth disease;
   II. Dipping the tail in hot water or by applying powdered camphor for overcoming tail neurosis;
   III. Feeding ground neem leaves for internal parasites;
   IV. Feeding sprouted whole wheat for 10-15 days continuously for anoestrus, etc.

The findings portray that a lot of wisdom still exists in rural areas but there is uncertainty about their expert validation.

Conclusion

ITK is still an under-utilized resource in the development activities. It needs to be intensively and extensively studied and incorporated into formal research and extension practices in order to make rural development strategies more sustainable. Special efforts are needed to understand, document and disseminate ITK for preservation, transfer or adoption elsewhere. We conclude that IK is used in agriculture, engineering, medicine, soil conservation and in many other fields. For instance, wooden hand hoes out of wood were used for cultivation, farmers knew which trees to get herbs from, fresh foods were obtained from the wilderness and people observed changes in climate by watching the entire environment. Intercropping is believed to increase on the crop yield per unit area and also to replenish the soil. Long periods of fallowing land were observed, mulching was practiced and crop rotation was equally important. Management strategies employed by the local people to exploit the environment show that there is a store of indigenous knowledge which people have developed over generations through daily observations and practices.

Indigenous knowledge still plays an important role in traditional farming in the culture, thus supporting local food security. The NPLM case is an excellent example of a community which is still practicing indigenous knowledge, and has higher food crop biodiversity. The potential food insecurity is not primarily due to decreasing yields, but more so because of the uncertainty of market prices. To protect biodiversity and ensure the in situ management of natural resources, an adequate farmer's income is required.

References

Dr. K. Vijayalakshmi and Subhashini Sridhar, CIKS Chennai. Pesticide Monitor Volume 8, No. 3, October 1999.

Role of ITK in Conservation Agriculture: Blending Indigenous and Scientific Knowledge. N. Prakash, S. S. Roy and S. V. Ngachan. ICAR Research Complex for NEH Region, Umiam – 793103, Meghalaya

Indigenous Traditional Knowledge . Dr. Pranab Kumar Ghosh Ms. Bijoylaxmi Sahoo. Orissa Review January -2011 Members of Orissa Secretariat Data Entry Operators' Association greeting with Shri Bijoy Kumar Patnaik, Chief Secretary on New Year's Day at Secretariat on 1.1.2011.

Handbook of Agriculture, published by ICAR.

CHAPTER 6

Crop Voyage in India and World

Crops–Indigenous and Introduced

Since time immemorial, cereals, particularly wheat, rice, and maize are considered to be life sustaining crops for humans. These crops will play a pivotal role in food security system of several nations across the world in future. The utilization of cereals as food and feed, and for industrial purpose is around 1792 million out of which wheat, rice, and coarse grains contribute nearly 35.4%, 20.8%, and 53.7% respectively. In the past fifty years the world has witnessed structural change in cereal economics:

Long run trend towards wheat and rice and to some extent for maize, while replacement of coarse grain crops occurred. Developing countries achieved higher growth in production and consumption and at the same time recorded rise in deficits. Rapid expansion of cereals as feed in developing countries and increased share of cereals in world trade.

It has been projected that, in the next 25 years, world demand for cereals will increase by 2-3% per annum mainly due to increase in population as well as change in choice and income of the people. This growth rate clearly suggests that the world community is still facing the daunting task to maintain adequate food supply for larger sections of population in future.

History of agricultural development and food production.

Agricultural system	Cultural stage/time	World Population (million)	Land holding (ha person[1])
Hunting and gathering	Paleolithic	7	-
Shifting agriculture	Neolithic (10000 years ago)	35	40.0
Medieval rotation	500-1450 AD	900	1.5
Livestock farming	Late 1700s	1800	0.7
Improved farming	20th century	4200	0.3

[1]. Use of fertilizers, pesticides, and improved crop varieties. Source: Tisdale et al. (1993).

At present context, the concept of "sustainable agriculture development" has been introduced to the world community by the "Bruntland Commission", which has been accepted by all the countries. It ensures that"long-term effects of development do not damage the rightful heritage of future generations." More specifically,it is defined as "increasing production to satisfy growing demands while at the same time preserving basic ecological integrity of production system." Such sustainable development in agriculture emphasizes conservation of land, water, plant, and animal genetic resources through technically appropriate, economically viable, socially acceptable methods. Presently, utilization and access to genetic resources have become an increasingly important issue because:

I. Technological advancement in molecular biology and biotechnology have opened new avenues for widening gene pool of several crops.
II. Legal situation concerning ownership of genetic resources have changed (patent law, state ownership)
III. The decline/less recognition in value of biological diversity has made risk of their extinction/shortages.

Agricultural development in the recent past has markedly accelerated erosion of plant genetic material, loss of genetic diversity, or heterogeneity on one hand while on the other hand it has increased uniformity and genetic vulnerability of cultivated species to diseases and pests. This necessitates indepth knowledge of the history of cereals involving the origin, process of domestication, and morpho-physiological changes that occurred during the evolutionary process. These aspects can help in conservation of primitive types and further improvement in genetic capabilities of these crops.

Centers of Origin of Crop Plants

The origin of crop plants is now basic to plant breeding in order to locate wild relatives, related species, and new genes (especially dominant genes, sources of disease resistance). Knowledge of the origins of crop plants is vitally important in order to avoid genetic erosion, the loss of germplasm due to the loss of ecotypes and landraces, loss of habitat (such as rainforests), and increased urbanization. Germplasm preservation is done through gene banks (largely seed collections but now frozen stem sections) and preservation of natural habitats (especially in centers of origin).

The Eight Vavilovian Centers

Old World

I. Chinese Center: This largest independent center includes the mountainous regions of central and western China, and adjacent lowlands. A total of 136 endemic plants are listed, out of which, a few are known to us as important crops.

Cereals and Legumes

1. Broomcorn millet -*Panicum miliaceum*

2. Italian millet - *Panicum italicum*
3. Japanese barnyard millet -*Panicum frumentaceum*
4. Kaoliang -*Andropogon sorghum*
5. Buckwheat -*Fagopyrum esculentum*
6. Hull-less barley -*Hordeum hexastichum*
7. Soybean -*Glycine max*
8. Adzuki bean -*Phaseolus angularis*
9. Velvet bean -*Stizolobium hassjoo*

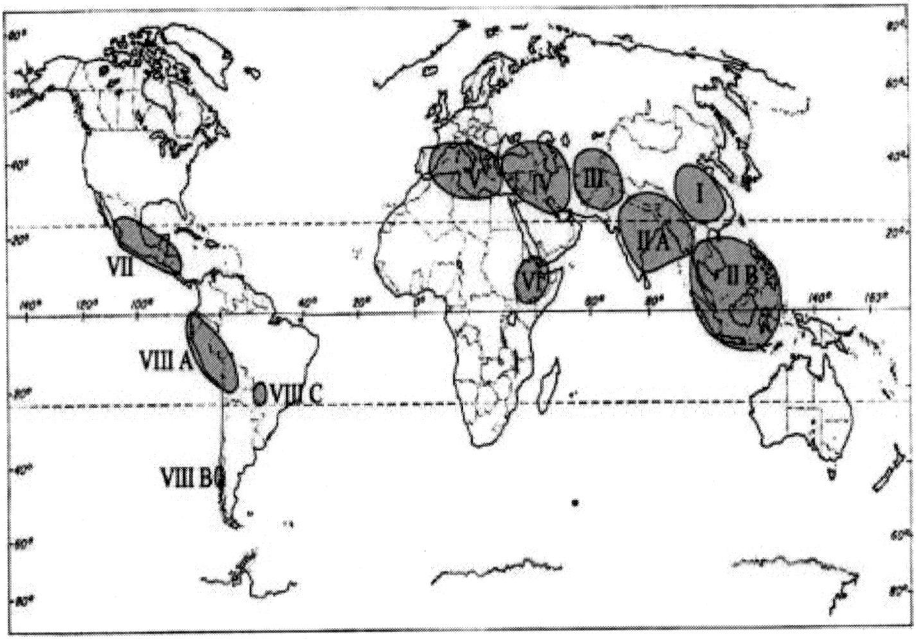

Fig. 6.1. *The eight Vavilovian centres of origin for crop plants*

Table – 6.2: List of crops originated in Chinese centre.

Sl. No	Near East	Americas	Far East
1	Lentil	Pink bean	Adsuki bean
2	Chickpea	Lima bean	Yellow banana
3	Salt	Manioc	Red banana
4	Pea	Potato	Green banana
5	Raisin	Summer squash	Soybean
6	Olive	Acorn squash	Coconut

7	Barley	Small dried chili pepper	Millet
8	Walnut	Fresh chili pepper	Yam
9	Almond	Corn (maize)	Sugar cane
10	Pistachio nut	Dried chili pepper	Rice
11	Apricot	Cocoa bean	
12	Date	Whole dried corn	
13	Wheat	Cracked dried corn	
14	Fig	Pinto bean	
15	Fava bean	Shelled pumpkin seed	

Roots, Tubers, and Vegetables

1. Chinese yam -*Dioscorea batatas*
2. Radish - *Raphanus sativus*
3. Chinese cabbage -*Brassica chinensis, B. pekinensis*
4. Onion -*Allium chinense, A. fistulosum, A. pekinense*
5. Cucumber - *Cucumis sativus*

Fruits and Nuts

1. Pear, Pyrus serotina -*P. ussuriensis*
2. Chinese apple -*Malus asiatica*
3. Peach -*Prunus persica*
4. Apricot -*Prunus armeniaca*
5. Cherry -*Prunus pseudocerasus*
6. Walnut -*Juglans sinensis*
7. Litchi -*Litchi chinensis*

Sugar, Drug, and Fiber Plants

1. Sugarcane -*Saccharum sinense*
2. Opium poppy -*Papaver somniferum*
3. Ginseng -*Panax ginseng*
4. Camphor -*Cinnamomum camphora*
5. Hemp -*Cannabis sativa*

II. Indian Center: This center has two subcenters

A. Main Center (Hindustan): It includes Assam and Burma, but not Northwest India, Punjab, nor Northwest Frontier Provinces. In this zone, 117 plants were considered to be originated.

Cereals and Legumes

1. Rice -*Oryza sativa*
2. Chickpea or gram -*Cicer arietinum*
3. Pigeon pea -*Cajanus indicus*
4. Urd bean -*Phaseolus mungo*
5. Mung bean -*Phaseolus aureus*
6. Rice bean -*Phaseolus calcaratus*
7. Cowpea -*Vigna sinensis*

Vegetables and Tubers

1. Eggplant -*Solanum melongena*
2. Cucumber -*Cucumis sativus*
3. Radish- *Raphanus caudatus* (pods eaten)
4. Taro -*Colocasia antiquorum*
5. Yam -*Dioscorea alata*

Fruits

1. Mango -*Mangifera indica*
2. Orange -*Citrus sinensis*
3. Tangerine -*Citrus nobilis*
4. Citron -*Citrus medica*
5. Tamarind -*Tamarindus indica*

Sugar, Oil, and Fiber Plants

1. Sugar cane -*Saccharum officinarum*
2. Coconut palm -*Cocos nucifera*
3. Sesame -*Sesamum indicum*
4. Safflower -*Carthamus tinctorius*
5. Tree cotton -*Gossypium arboreum*
6. Oriental cotton -*Gossypium nanking*
7. Jute -*Corchorus capsularis*
8. Crotalaria -*Crotalaria juncea*
9. Kenaf -*Hibiscus cannabinus*

Spices, Stimulants, Dyes, and Miscellaneous

1. Hemp -*Cannabis indica*
2. Black pepper -*Piper nigrum*
3. Gum arabic -*Acacia arabica*
4. Sandalwood -*Santalum album*
5. Indigo -*Indigofera tinctoria*

6. Cinnamon tree*Cinnamomum zeylanticum*
7. Croton -*Croton tiglium*
8. Bamboo -*Bambusa tulda*

B. Indo-Malayan Center: This centre includes Indo-China and the Malay Archipelago. Fifty-five plants were listed, including:

Cereals and Legumes

1. Job's tears -*Coix lacryma*
2. Velvet bean -*Mucuna utilis*

Fruits

1. Pummelo -*Citrus grandis*
2. Banana -*Musa cavendishii, M. paradisiaca, H. sapientum*
3. Breadfruit -*Artocarpus communis*
4. Mangosteen -*Garcinia mangostana*

Oil, Sugar, Spice, and Fiber Plants

1. Candlenut -*Aleurites moluccana*
2. Coconut palm -*Cocos nucifera*
3. Sugarcane -*Saccharum officinarum*
4. Clove -*Caryophyllus aromaticus*
5. Nutmeg -*Myristaca fragrans*
6. Black pepper -*Piper nigrum*
7. Manila hemp or abaca -*Musa textilis*

III. Central Asiatic Center: Includes Northwest India (Punjab, Northwest Frontier Provinces and Kashmir), Afghanistan, Tadjikistan, Uzbekistan, and western Tian-Shan. Forty-three plants are listed for this center, including many wheats.

Grains and Legumes

1. Common wheat -*Triticum vulgare*
2. Club wheat -*Triticum compactum*
3. Shot wheat -*Triticum sphaerocoecum*
4. Pea -*Pisum sativum*
5. Lentil -*Lens esculenta*
6. Horse bean -*Vicia faba*
7. Chickpea -*Cicer arientinum*
8. Mung bean -*Phaseolus aureus*
9. Mustard -*Brassica juncea*
10. Flax -*Linum usitatissimum* (one of the centers)
11. Sesame -*Sesamum indicum*

Fiber Plants

1. Hemp -*Cannabis indica*
2. Cotton -*Gossypium herbaceum*

Vegetables

1. Onion -*Allium cepa*
2. Garlic -*Allium sativum*
3. Spinach -*Spinacia oleracea*
4. Carrot -*Daucus carota*

Fruits

1. Pistacia -*Pistacia vera*
2. Pear -*Pyrus communis*
3. Almond -*Amygdalus communis*
4. Grape -*Vitis vinifera*
5. Apple -*Malus pumila*

IV. Near-Eastern Center: This centre includes interior of Asia Minor, all of Transcaucasia, Iran, and the highlands of Turkmenistan. 83 species including nine species of wheat were listed for this region.

Grains and Legumes

1. Einkorn wheat -*Triticum monococcum* (14 chromosomes)
2. Durum wheat -*Triticum durum* (28 chromosomes)
3. Poulard wheat -*Triticum turgidum* (28 chromosomes)
4. Common wheat, -*Triticum vulgare* (42 chromosomes)
5. Oriental wheat -*Triticum orientale*
6. Persian wheat -*Triticum persicum* (28 chromosomes)
7. *Triticum timopheevi* (28 chromosomes)
8. *Triticum macha* (42 chromosomes)
9. *Triticum vavilovianum*, branched (42 chromosomes)
10. Two-row barleys -*Hordeum distichum, H. nutans*
11. Rye -*Secale cereale*
12. Mediterranean oats -*Avena byzantina*
13. Common oats -*Avena sativa*
14. Lentil -*Lens esculenta*
15. Lupine -*Lupinus pilosus, L. albus*

Forage Plants

1. Alfalfa -*Medicago sativa*
2. Persian clover -*Trifolium resupinatum*

3. Fenugreek -*Trigonella foenum graecum*
4. Vetch -*Vicia sativa*
5. Hairy vetch - *Vicia villosa*

Fruits

1. Fig -*Ficus carica*
2. Pomegranate -*Punica granatum*
3. Apple -*Malus pumilo* (one of the centers)
4. Pear -*Pyrus communis* and others
5. Quince -*Cydonia oblonga*
6. Cherry -*Prunus cerasus*
7. Hawthorn -*Crataegus azarolus*

V. Mediterranean Center: This zone includes the borders of the Mediterranean Sea. 84 plants are located in this region including olive and many cultivated vegetables and forages.

Cereals and Legumes

1. Durum wheat -*Triticum durum expansum*
2. Emmer -*Triticum dicoccum* (one of the centers)
3. Polish wheat -*Triticum polonicum*
4. Spelt -*Triticum spelta*
5. Mediterranean oats -*Avena byzantina*
6. Sand oats -*Avena brevis*
7. Canarygrass -*Phalaris canariensis*
8. Grass pea -*Lathyrus sativus*
9. Pea -*Pisum sativum* (large seeded varieties)
10. Lupine -*Lupinus albus,* and others

Forage Plants

1. Egyptian clover -*Trifolium alexandrinum*
2. White Clover -*Trifolium repens*
3. Crimson clover -*Trifolium incarnatum*
4. Serradella -*Ornithopus sativus*

Oil and Fiber Plants

1. Flax -*Linum usitatissimum*, and wild *L. angustifolium*
2. Rape -*Brassica napus*
3. Black mustard -*Brassica nigra*
4. Olive -*Olea europaea*

Vegetables

1. Garden beet -*Beta vulgaris*

2. Cabbage -*Brassica oleracea*
3. Turnip -*Brassica campestris, B. napus*
4. Lettuce -*Lactuca sativa*
5. Asparagus -*Asparagus officinalis*
6. Celery -*Apium graveolens*
7. Chicory -*Cichorium intybus*
8. Parsnip -*Pastinaca sativa*
9. Rhubarb -*Rheum officinale*

Ethereal Oil and Spice Plants

1. Caraway -*Carum carvi*
2. Anise -*Pimpinella anisum*
3. Thyme -*Thymus vulgaris*
4. Peppermint -*Mentha piperita*
5. Sage -*Salvia officinalis*
6. Hop -*Humulus lupulus*

VI. Abyssinian Center: This includes Abyssinia, Eritrea, and part of Somaliland. In this center, 38 species were listed including wheat and barley.

Grains and Legumes

1. Abyssinian hard wheat -*Triticum durum abyssinicum*
2. Poulard wheat -*Triticum turgidum abyssinicum*
3. Emmer -*Triticum dicoccum abyssinicum*
4. Polish wheat -*Triticum polonicum abyssinicum*
5. Barley -*Hordeum sativum* (great diversity of forms)
6. Grain sorghum -*Andropogon sorghum*
7. Pearl millet -*Pennisetum spicatum*
8. African millet -*Eleusine coracana*
9. Cowpea -*Vigna sinensis*
10. Flax -*Linum usitatissimum*

Miscellaneous

1. Sesame -*Sesamum indicum* (basic center)
2. Castor bean -*Ricinus communis* (a center)
3. Garden cress -*Lepidium sativum*
4. Coffee -*Coffea arabica*
5. Okra -*Hibiscus esculentus*
6. Myrrh -*Commiphora abyssinicia*
7. Indigo -*Indigofera argente*

New World

VII. South Mexican and Central American Central: This centre includes southern sections of Mexico, Guatemala, Honduras and Costa Rica.

Grains and Legumes

1. Maize -*Zea mays*
2. Common bean -*Phaseolus vulgaris*
3. Lima bean -*Phaseolus lunatus*
4. Tepary bean -*Phaseolus acutifolius*
5. Jack bean -*Canavalia ensiformis*
6. Grain amaranth -*Amaranthus paniculatus leucocarpus*

Melon Plants

1. Malabar gourd -*Cucurbita ficifolia*
2. Winter pumpkin -*Cucurbita moshata*
3. Chayote -*Sechium edule*

Fibre Plants

1. Upland cotton -*Gossypium hirsutum*
2. Bourbon cotton -*Gossypium purpurascens*
3. Chayote -*Sechium edule*

Miscellaneous

1. Sweetpotato -*Ipomea batatas*
2. Arrowroot -*Maranta arundinacea*
3. Pepper -*Capsicum annuum, C. frutescens*
4. Papaya -*Carica papaya*
5. Guava -*Psidium guayava*
6. Cashew -*Anacardium occidentale*
7. Wild black cherry -*Prunus serotina*
8. Cochenial -*Nopalea coccinellifera*
9. Cherry tomato -*Lycopersicum cerasiforme*
10. Cacao -*Theobroma cacao*
11. Tobacco -*Nicotiana rustica*

VIII. South American Center: (62 plants listed) In this centre, three subcenters are found.

Peruvian, Ecuadorean, Bolivian Center: This centre comprised of the high mountainous areas, formerly the center of the Megalithic or Pre-Inca civilization.

Endemic plants of the Puna and Sierra high elevation districts included:

Root Tubers

1. Andean potato -*Solanum andigenum* (96 chromosomes)
2. Other endemic cultivated potato species. Fourteen or more species with chromosome numbers varying from 24 to 60.
3. Edible nasturtium -*Tropaeolum tuberosum.*
4. Coastal regions of Peru and non-irrigated subtropical and tropical regions of Ecuador, Peru and Bolivia included:

Grains and Legumes

1. Starchy maize -*Zea mays amylacea*
2. Lima bean -*Phaseolus lunatus* (secondary center)
3. Common bean -*Phaseolus vulgaris* (secondary center)

Root Tubers

1. Edible canna -*Canna edulis*
2. Potato -*Solanum phureja* (24 chromosomes)

Vegetable Crops

1. Pepino -*Solanum muricatum*
2. Tomato -*Lycopersicum esculentum*
3. Ground cherry -*Physalis peruviana*
4. Pumpkin -*Cucurbita maxima*
5. Pepper -*Capsicum frutescens*

Fiber Plants

1. Egyptian cotton -*Gossypium barbadense*

Fruit and Miscellaneous

1. Passion flower -*Passiflora ligularis*
2. Guava -*Psidium guajava*
3. Heilborn -*Carica candamarcensis*
4. Quinine tree -*Cinchona calisaya*
5. Tobacco -*Nicotiana tabacum*

B. Chiloe Center (Island near the coast of southern Chile)

1. Common potato -*Solanum tubersum* (48 chromosomes)
2. Wild strawberry -*Fragaria chiloensis*

C. Brazilian-Paraguayan Center

1. Manioc -*Manihot utilissima*
2. Peanut -*Arachis hypogaea*
3. Rubber tree -*Hevea brasiliensis*
4. Pineapple -*Ananas comosa*

5. Brazil nut -*Bertholletia excelsa*
6. Cashew -*Anacardium occidentale*
7. Purple granadilla -*Passiflora edulis*

History of Rice, Sugarcane and Cotton

Rice

Rice supplies a quarter of the entire caloric intake of the human and therefore, it is the most important tropical cereal crop. South and Southeast Asia contribute about 90% of its area and consumption, and this areas support a major part of the world population. Rice belongs to the genus *Oryza* and there are two main ecotypes, i.e., *sativa* in Asia and *glaberrima* in Africa. Rice is a semi aquatic graminaceous crop and it can grow in complex range of environments, i.e., from uplands at altitude of 3000m to rainfed lowland irrigated, tidal swamp, and deepwater areas. Besides these two species, aquatic rice species, i.e., *Zizania aquatica* and *Z. palustris,* are endemic to North America, where it is the staple food of Indians.

Origin

The place of major diversity where rice might have domesticated is roughly the east west belt along the Himalayas and adjoining Asia mainland (from Assam, Bangladesh, Burma, Thailand, southern China, and northeren Vietnam). The archaeological evidence suggests that Asian rice culture was established around 7000 years ago. Carbonized grains excavated from Hastinapur (New Delhi) suggested that rice was in cultivation during 1100-800 BC. Subsequently, the grain samples collected at Atrankikar (Uttar Pradesh) were considered as oldest (1500-1100 BC). It has been inferred from the excavation of rice samples belonging to 5000-4000 BC in Thailand that from this place rice spread to other countries.

Evolutionary History

The evidences from diverse disciplines including biosystematic and paleogeology suggest that the genus Oryza arose from a common ancestor. The evolutionary path was from wild perennial to wild annual to cultivated annual, and the closely related wild relatives contributed differentiation of two cultigens. In *Oryza sativa,* the evolution of different geographical races, i.e., japonica, javanica, and indica, took place assisted strongly by human selection. It has been believed that both Asia and Africa elongation and floating ability in two cultigens was derived from their wild ancestor. Regarding development ofperennial to annual types, a theory has been suggested that climatic changes during the Pleistocene period induced physiological stress in herbaceous flora, which lead to evolution of annuals from perennials. In rice, change might have occurred in the following sequence:

Perennial->climatic stress->seasonal->human selection-> cultivated rice.

Future Strategies

The primitive cultivars and allied wild species of cultivated rice constitute a store house of rare and valuable genes but their use in the breeding programme is limited because these having many undesirable characters such as shattering of grains, sterility, and red grains. In recent years, the precise elimination of undesirable characters while maintaining vital characteristics has been found possible as evident from the fact that:

Four different spontanea types (*O.rufipogon*) subjected to different chemical mutagens treatment, produced short plants with thick, erect leaves, and profuse tillering and possessed a different source of dwarfening gene.

Short culm mutants induced in the Assam type of *O. rufipogon* produced higher yield and proved more adaptable under waterlogged conditions during the monsoon compared to the variety IR 8. Rice species growing under marshy areas provide excellent sources of resistance to drought and waterlogged conditions.

Agri-history of Cotton in India: An Overview

The antiquity of cotton in the Indian subcontinent has been traced to the 4th millenium BC. The fabrics, recovered from the Mohenjo-daro excavations in Sind (Pakistan), during approximately 3000 BC, were identified to have originated from cotton plants, closely related to the *Gossypium arboreum* species. The lint-bearing species of the genus *Gossypium*, the true cottons, are four, out of which the diploid (2n=26) species *G. arboreum* and *G. herbaceum* are considered to originated in Asia and Africa.

The history of introduction into India of the new world cottons (tetraploid species of *G. hirsutum* and *G. barbadense* with 2n=52) held back to the 18th century AD. By the last decade of the 20th century, India had gained 1st place in the global cotton statistics with the largest cropped area of 8.9 million in 1996-97, but the productivity is still to be improved. India also is growing the most diverse cultivars of cotton in terms of botanical species and composition, producing the widest range of cotton fibre quality suitable for spinning 6's to 120's counts yarn, and supporting the largest agro-based national industry of the country.

Origin of the Indigenous Cottons

The cotton textiles of the Harappan civilization (2300-1750BC) were produced by sophisticated textile craftsmanship. The earliest agricultural discovered suggested that true cottons were already present in the Indian subcontinent. Wild and weedy types have been found to be associated with primitive cultivated types in both the old world species of *G. herbaceum* and *G. arboreum*.

Species of *G. herbaceum*, have been found from the coastal strip northwest of Karachi (Pakistan), through northern Baluchistan to south Yemen, Ethiopia, and Sudan and even in West Africa south of the Sahara. Species of *G. arboreum*. have been recorded by in Kathiawar, Gujarat, Khandesh, and the Deccan in India. Differentiation of the three perennial races of *G. arboreum,* namely *burmanicum* of northeastern India, *indicum* of western India and the Penninsula, and *sudanense* of northern Africa, ante-dated domestication and each contributed separately to the cultivated cottons in Asia and Africa.

Agri-history of Cotton Production Development

Until the middle of the 18th century, only indigenous *arboreum* and *herbaceum* varieties of cotton were grown in different regions of the country. Due to the human skills and technology of the local artisans, very fine yarns were produced by them, even from the short staple and coarse cottons grown in India. In 1788, the Governor General (at Calcutta) was requested by London to encourage growth and improvement of Indian cottons to meet the requirements of the Lancashire textile industry.

Sugarcane

The origin of sugarcane was India. The species *Saccharum officinarum* was first domesticated in India and the spread to other countries by Arab merchants. Evidences revealed that 3000-7000 years ago, Atarna veda indicated that sugarcane originated from the Sakkaram area and later in Sanskirit, it was indicated as sakkra. Earlier indications in Kautilya Artha Sastra also mentioned about the cowdung sett treatment for sugarcane.

List of major plant species domesticated, introduced and cultivated in India.

Crops domesticated in Indian sub-continent	
Cereals	*Oryza sativa* (Rice)
Millets and Forages	*Cenhrus ciliaris* (bunch grass), *Coix lacrymz-jobi* (job's tears), *Digitaria cruciata* (Digitaria), *Echinochloa colona* (Deccan grass), *Echinochloa crus-galli* (cockspur grass), *Panicum arntidotale* (blue panicum), *Panicum miliaceum* (french millet), *Paspalum scrobiculatum* (kodo-millet), *Panicum sumatrense* (little millet), *Sesbania bispinosa syn. Sesbania aculeata* (dhaincha)
Grain legumes	*Cajanus cajan* (pigeonpea), *Dolichos biflorus syn. Dolichos lablab* (hyacinth bean), *Macrotyloma uniflorum* (horse gram/kulthi), *Psophocarpus tetragonolobus* (Goa bean), *Mucuna utilis* (velvet bean), *Vigna radiata* (green gram), *Vigna aconitifolia* (moth bean), *Vigna mungo* (black gram), *Vigna umbellata* (rice bean),*Vigna angularis* (adzuki bean), *Vigna triloba* (jungi bean)
Oilseeds	*Brassica juncea* (Indian mustard), *Brassica rapa var. toria* (Indian rape), *Brassica rapa ssp. trilocularis* (yellow sarson or Indian Colza), *Brassica rapa var. dichotoma* (brown sarson), *Sesamum indicum* (sesame)
Fibre crops	*Corchorus capsularis* (white jute), *Corchorus olitorius* (tossa jute), *Crotolaria juncea* (sun- hemp), *Agave cantala* (kantala), *Bombax malabaricum* (red silk cotton), *Gossypium arboreum* (tree-cotton), *Gossypium obustifolium* (Asiatic cotton), *Hibiscus sabdariffa* (Jamaica sorrel),*Hibiscus cannabinus* (Deccan hemp).

Vegetables	*Amaranthus tricolour* (lal sag), *Amaranthus blitum* (sag chulai), *Abelmoschus esculentus* (okra),*Amorphophallus campanulatus* (elephant-foot yam), *Chenopodium album* (common pigwheat), *Colocasia esculenta* (taro), *Coccinia indica* (kovai fruit), *Cucumis sativus* (cucumber), (ridge gourd), *Luffa cylindrica* (sponge gourd), *Momordica charantia* (bitter gourd), *Momoridica dioica* (balsam apple), *Momordic cochinchinesis* (kakora), *Moringa oleifera* (drumstick), *Praecitrullus fistulosus* (round gourd), *Rumex vesicarius* (bladder dock),*Raphanus caudatus* (rat-tail Radish), *Solanum melongena* (brinjal), *Sesbania grandiflora* (agathi), *Spinacia oleracea* (spinach), *Trichosanthes anguina* (snake gourd), *Trichosoanthes dioica* (pointed gourd)
Fruits	*Aegle marmelos* (Bengal quince), *Areca catechu* (areca nut), *Artocarpus heteophyllus* (jack fruit). *Citrus limon* (lemon), *Carissa congesta* (karaunda), *Mangifera indica* (mango), *Musa paradisiaca* (plantain/banana), *Phoenix sylvestris* (date sugar palm, Indian palm), *Zizyphus nummularia* (wild jujube), *Zizyphus jujuba* (Indian jujube).
Some important medicinal and aromatic plants	*Abelmoschus moschatus* (muskmallow), *Atropa acuminata* (Indian belladona), *Azadirachta indica* (margosa tree), *cassia fistula* (Indian laburnum), *Cymbopogon martini* (palmarosa), *Cymbopogon nardus* (citronella grass), *Cymbopogon pendulus* (lemon grass), *Cynodon dactylon* (Bermuda grass), *Datura metel* (datura), *Emblica officinale* (Indian gooseberry), *Patchouli* (patchouli), *Rauvolfia serpentina* (serpentine root), (costus), *Vetiveria zizanioides* (vetiver)
Narcotics Spices and condiments	*Cannabis sativa* (hemp) *Cinnamomum tamale* (Indian cassis) *Cinnamomum verum* (cinnamon), *Crocus sativus* (saffron), *Curcuma amada* (mango ginger), *Curcuma caesia* (black turmeric), *Curcuma longa* (turmeric), *C. domestica syn. Curcuma zedoaria* (zedoary), *Elettaria cardamomum* (small cardamom), *Kaempferia galanga* (chandramula), *Mentha piperita* (mint), *Murraya koenigii*
Spices and Condiments	(curry leaf tree), *Myristica malabaricum* (nutmeg), *Piper betel* (betel pepper), *Piper longum* (long pepper), *piper nigrum* (black pepper), *Trigonella foenum-graecum* (fenugreek), *Zingiber officinale* (ginger)
Others	*Acacia nilotica* (India acacia), *Acacia catechu* (kattha), *Bambusa tulda* (Bengal bamboo), *Bambusa arundinacea* (thorny/spiny bamboo), *Bauhinia purpurea* (camel foot tree), *Borassus flabellifer* (palmyra-palm), *Camellia sinensis var. assamica* (tea), *Caesalpinia sappan* (Indian redwood), *Cedrela toona* (red cedar), *Cordia myxa* (Indian cherry), *Coffea bengalensis* (coffee), *Curcum angustifolia* (Indian arrowroot), *Dendrocalamus hamiltonii* (Dendrocalamus), *Dendrocalamus strictus* (Calcutta bamboo), *Dioscorea alata* (greater yam), *Dioscorea esculenta* (lesser yam), *Ficus elastica* (Indian rubber), *Ficus bengalensis* (banyan tree), *Ficus religiosa* (peepal), *Garcinia sylvestris* (wild mangosteen), *Indigofera tinctoria* (indigo), *kochia indica* (bui), *Lawsonia alba* (henna), *Marsdenia tinctoria* (rion), *Maoutia puya* (pua, poi), *Morinda angustifolia* (ban haldi), *Morinda citrifolia* (Indian mulberry), *Nephelium longana* (anshaphal), *Nerium indicum* (kaner), *Nyctanthes arbortristis* (Tree of Sadness), *Ochlandra travancorica* (elephant grass), *Pluchea indica* (pludina), *Oroxylum indicum* (sonapatha), *Rubia cordifolia* (Indian madder), *Saccharum sinense* (sugarcane), *Saccharum officinarum* (sugarcane), *Sapindus trifoliatus* (soap nut tree), *Sida rhombifolia* (Cuba jute), *Sinocalamus giganteus* (Sinocalamus), *Tamarindus indica* (tamarind)

Crops Introduced in India by Different Nations

Crops Introduced by Portuguese

Oilseeds -*Arachis hypogeae* (groundnut)

Pseudo cereals -*Amaranthus caudatus* (amaranth)

Vegetables -*Cucurbita moschata* (pumpkin), *Solanum tuberosum* (potato) *Ipomoea batatas* (sweet potato),

Fruits - *Anacardium occidentale* (cashew nut), *Anona squamosa* (custard apple), *Psidium guajava* (guava)

Spices and condiments -*Capsicum annuum* (chilli)

Narcotics -*Nicotiana tabaccum* (tobacco)

Crops Introduced by Britishers

Grain legumes - *Castanospermum australe* (black bean), *Pisum sativum* (pea)

Pseudo cereals- *Avena sativa* (oat)

Fibre crops - *Gossypium barbadense* (cotton)

Vegetables -*Allium tuberosum* (leek), *Asparagus racemosus* (satawar), *Beta vulgaris* (beet root), *Brassica oleracea var. gemmifer* (Brussels), *Brassica oleracea var. botrytis* (cauliflower), *Capsicum frutescens* (sweet pepper), *Cichorium intybus* (chicory), *Cucurbita maxima* (squash), *Daucas carota* (carrot, orange type), *Lactuca sativa* (lettuce), *Lycopersicon esculentum* (tomato), *Pisum sativum* (sweet pea)

Fruits - *Averrhoa bilimbi* (bilimbi), *Averrohoa carambola* (carambola), *Carica papaya* (papaya), *Eugenia jambos* (rose apple), *Fragaria ananassa* (strawberry), *Garcinia mangostana* (mangosteen), *Helianthus tuberosus* (artichoke), *Manihot esculenta* (cassava), *Malus pumila* (apple), *Prunus avium* (cherry), *Prunus armeniaca* (apricot), *Prunus persica* (peach), *Prunus communis syn. P. domestica* (plum), *Pyrus communis* (pear), *Ribes rubrum* (red currant)

Medicinal and aromatic plants -*Cinchona officinalis* (quinine), *Origanum vulgare* (majoram), *Papaver somniferum* (opium poppy), *Pelargonium capaitatium* (Geranium), *Salvia officinalis* (sage), *Thymus vulgaris* (thyme), *Vanilla aromatica* (vanilla).

Crops introduced from West and Central Asia by Mughals or Arabs

Pulse - *Pisum sativum* (pea)

Vegetables - *Brassica rapa* (turmip), *Brassica oleracea var. capitata* (cabbage),*Daucas carota* (carrot, black & red type)

Spices and condiments - *Allium sativum* (garlic), *Allium cepa* (onion), *Coriandrum sativum* (coriander), *Syzygium aromaticum* (clove)

Fruits - *Cucumis melo* (sweet muskmelon), *Phoenix dactylifera* (date palm), *Vitis vinifera* (grape)

Crops Introduced by Spaniards

Vegetables - *Phaseolus vulgaris* (French bean)

Crops Introduced from China

Oilseed - *Aleurites fordii* (tung-oil), *Glycine max* (soyabean)

Fruits - *Juglans regia* (walnut), *Litchi chinensis* (litchi)

Others - *Eriobotrya japonica* (loquat), *Sapium sebiferum* (tallow-tree)

Crops Introduced from Latin America

Fruits - *Ananas comosus* (pineapple),

Plantation crop -*Hevea brasiliensis* (Rubber)

Crops Introduced from Southeast Asia and Pacific Islands

Fruits - *Arenga pinnata* (sugar-palm), *Artocarpus communis* (breadfruit), *Citrus paradisi* (grapefruit), *Citrus decumanus* (pomelo)

Others -*Durio zibethinus* (durian) and *Metroxylon sagus* (sago)

Some Recent Introductions

Humulus lupulus (hops), *Helianthus annuus* (sunflower), *Simarouba glauca* (simarouba), *Cyphomandra betacea* (tree tomato), *Carya illinoensis* (pecan nut), *Corylus avellana* (hazel nut), *Macadamia tetraphylla* (macadamia nut), *Parthenium argentatum* (guayule), and *Mentha arvensis* (spearmint, USA) *Acacia senegal* (Australia), *Acacia mangium* (Australia) and *Actinidia chinensis* (kiwifruit, New Zealand)

CHAPTER 7

Agriculture Scope, Importance of Agriculture and Agricultural Resources Available in India

Agriculture helps to meet the basic needs of human and animal by providing food, clothing, shelters, medicine and recreation. Agriculture is the most important sector in the world for their livelihood. Natural resource like land, light, air, temperature and rain water etc., are integrated into single primary unit indispensable for human beings. Secondary productive units like animals including livestock, birds and insects, feed on byproduct of agricultural crops and provide valuable products such as meat, milk, wool, eggs, honey, silk and lac. Agriculture provides food, feed, fibre, fuel, furniture, raw materials and materials for and from factories, provides fresh environment and abundant food for human beings. Satisfactory agricultural production brings peace, prosperity, harmony, health and wealth to individuals of a nation. Agriculture helps to elevate the community consisting of different castes and clauses, thus it leads to a better social, cultural, political, environmental and economical life.

Definition of Agriculture

1. The word agriculture comes from the Latin words ager, means the soil & cultura, means cultivation.
2. Agriculture can be defined as the cultivation and or production of crop plants or livestock products.
3. Agriculture is defined as the art, the science, and the business of producing crops and the livestock for economic purposes.
4. Agriculture includes Crop Production, Animal Husbandry and Dairy Science & Technology, Soil Science, Horticulture, Agricultural Economics, Agricultural Engineering, Botany, Plant Pathology, Entomology and Extension Education which develops its separate and distinct branches of agriculture.

Agriculture as a science: It utilizes all modern technologies developed on scientific principles such as crop improvement, crop production, crop protection, economics etc., to maximize the yield and profit. For example, new crops and varieties developed by hybridization, transgenic crop varieties, resistant to pests and diseases, high fertilizer responsive varieties, water management, herbicides to control weeds, drought resistant varieties and use of bio-control agents to combat pest and diseases etc.

Agriculture as an art: It embraces knowledge to perform the operations of the farm in a skillful manner. The skill is classified as:

Physical skill: It involves the ability and capacity to carry out the field and laboratory operation in an efficient way for e.g., handling of farm implements, animals etc., sowing of seeds, fertilizer and pesticides application etc.

Mental skill: The farmer is able to take a right decision based on experience, such as (i) time and method of ploughing, (ii) selection of crop and cropping system to suit soil and climate, (iii) adopting improved farm practices etc.

Agriculture as the business: As long as agriculture is the way of life of the rural peoples, production is ultimately bound to consumption and other uses. But agriculture as a business aims at maximum net return through the management of land, labour, water and capital, employing the knowledge of various agrotechnologies for production of food, feed, fibre and fuel. In recent years, agriculture is commercialized to run as a farm business through mechanization.

Scope of Agriculture in India

In India, human population is increasing while area under cultivation is decreasing, which demands intensification of cropping and allied activities in two dimensions *i.e.* time and space dimension. India occupied under tropical climate with abundant solar energy throughout the year, which favours growing crops round the year. There is a vast scope to increase irrigation potential by river projects, underground source and minor irrigation projects. In additional to the above, India's having availability of cheap wage. In India, food grain production increased almost four folds from about 50 million tons at independence to more than 264 million tons in 2013 to 2014 through green revolution. Despite variation in the performance of individual crops and regions, growth of total food grain production over previous year was 2.88 % which kept ahead of population growth at about 1.3 % per annum (2010 to 2015). Through the white revolution, milk production increased from 17million tones at independence to 132.4 million tons annually.

Through blue revolution, fish production rose from 0.75 million tones to nearly 9.6 million tons during the last six decades (2012–13). Through yellow revolution, oil seed production increased about 11 times (from 3 million tons to 32.4 million tones, 2013–2014) since independence. Similarly, the egg production increased from 2 billion at independence to 69.73 billion (2012–13), sugarcane production from 57 million tons to 354.95 million tones estimated (2014–15), cotton production from 3 million bales to 37.5 million bales (2013- 2014) which shows our sign of progress. India is the first producer of milk and second largest producer of fruits and vegetables in the world.

In future, agriculture development in India would be guided not only by the compulsion of improving food and nutritional security, but also by the concerns for environmental protection, sustainability and profitability with organic intervention. The General Agreement on Trade and Tariff (GATT) and the liberalization process, globalization of markets would call for competitiveness and efficiency of agricultural production with good quality. Agriculture will face challenging situations on the ecological, global climate, economic equity, energy and employment fronts in the coming years.

Importance of Agriculture in India

Agriculture is the backbone of the Indian economy and despite concerted industrialization in the last 50 years; agriculture still occupies a place of pride. Agriculture is contributing nearly 30 per cent of the national income, providing employment to about 70 per cent of the employment generation and accounting for a sizable share of the country's foreign exchange earnings. It provides the food grains to feed the large population of the country. It is also the supplier of raw material to many industries. Thus, the very economic structure of the country rests upon agriculture. The importance of agriculture in the Indian economy is elaborated below:

(a) **Contribution to national income:** The National Income Committee and the Central Statistical Organization clearly shows that agriculture contributed about 56 per cent of the national income in 1950–51 but contributed only 22 per cent of the national income in 2006–07. The share of agriculture in the national income was about 56 per cent during 1950–51 and remained above 50 per cent during the following 30 years. However, the contribution of agriculture has declined in the last 25 years due to rapid increase in the production of industrial goods and services. A comparison between shares of agriculture in national income in India with that in other countries further depict dominance of agriculture in the Indian economy.

(b) **Contribution to employment generation:** Majority of the Indian population is directly or indirectly involved in agriculture for their livelihood. The decennial censuses indicate that 70 per cent of the population is depending on agriculture. These censuses show that an overwhelming majority of farm workers have been engaged in cultivation. Dependence of working population on other fields of agriculture like livestock, fisheries, forest etc., is also significant. The rate and pattern of investment in other economic sectors have not been such as to draw away surplus rural laborer and relieve the pressure of population on land. Since the growth of agricultural sector was very slow it failed to create enough opportunities for additional employment and less profit. In backward and underdeveloped countries the proportion of working population engaged in agriculture sector is quite high. For instance, it is 42 per cent in Egypt, 52 per cent in Indonesia and 72 per cent in China.

(c) **Contribution of manpower to industrial sector:** The agricultural laborer of rural sector has been the supplier of manpower to industry. The Commission on laborer is indicative that the Indian factory operatives were nearly all migrants from the rural areas. This drift to urban areas continues because of higher income generation. This is due to lack of opportunities for employment and income in rural areas on the one hand and lure of employment, higher income and urban facilities on the other.

(d) **Contribution to foreign exchange of agricultural resources:** Agricultural products and manufactured through agricultural raw material occupy an important place in the country's export. According to an estimate, agricultural commodities like raw cotton and jute, unmanufactured tobacco, oilseeds, spices, tea and coffee accounted for about 50 per cent of the total value of exports. This makes a remarkable contribution to the foreign exchange of the country.

(e) **Interdependence between agriculture and industry:** This is to the supply of raw materials and inputs from agriculture to industry and vice-versa. The supply of wage goods to the industrial sector, The supply of basic consumption goods to the agricultural population, the supply of materials for the building up of economic and social overheads in the agricultural sector. The interdependence between agriculture and industry is becoming stronger as the economy is developing. The application of science and technology in agriculture induces innovations in respect of industrial products, which are used for agricultural production. Agricultural inputs like fertilizers, pesticides, diesel oil, electric motor, diesel engines, pump sets, agricultural tools and implements, tractors, power tillers etc., are supplied by the industry and oil, sugar, jute and cotton textiles and tobacco industries rely heavily on the agricultural industry sector. The processing industries, which are utilizing agricultural raw material viz. fruit canning, milk products and meat products.

(f) **Contribution to Assets formation:** The pace of development is largely determined by the rate at which production assets increase. Before independence, the capital formation in Indian agriculture was of a low order and agriculture suffered from constant low yield technology, inequitable land tenure system and exploitation of the rural masses. The capital formation includes land development, construction of houses etc. Since independence, much more investment both public and private has been made in agriculture sector. The creation of physical assets has generally taken the form of land development, construction of irrigation facilities, roads and communication, farm buildings, agricultural machinery and equipment, warehouses, cold storages, market yard etc. This capital formation is helping not only development of agriculture but also the entire economy of the nation.

(g) **Contribution to purchasing power of farmers:** Agriculture provides purchasing power not only to those directly engaged in it but to others also who are in the industries and services. New opportunities for hundreds of blacksmiths, carpenters, masons, weavers, potters, leather workers, utensil-makers, tailors, cotton ginners, oil pressers, transporters and countless others. Thus, there are many industries, the prosperity and employment of which are dependent upon the purchasing power of the agricultural population or entrepreneurs. Hence, it is concluded that besides purchasing food for non-agricultural workers and raw materials for consumer industries, it has created demands for a great many new industries, which, in turn, have provided high and well paid employment. This existing role of agriculture in the Indian economy points out the necessity for the development of Indian agriculture to the fullest extent possible as the prosperity of agriculture largely stands for the prosperity of the Indian economy. The significance of agriculture lies in the fact that the development in agriculture is an essential condition for the development of the Indian economy.

Agricultural Resources Available in India

Agriculture has been the backbone of the Indian economy and it will continue to remain so for a long time as India have a full of natural resources. It has to support almost 17 per cent of world population from 2.3 per cent of world geographical area and 4.2 per cent of world's water resources existing here. Presently, the total net irrigated area covers 45.5 per cent of the net sown area; the remaining 54.5 per cent is rainfed conditions. The degradation of land and surface as well as ground water resources results in fast deterioration of soil health resulting less production. Losses due to biotic (insect-pests, diseases, weeds) and abiotic (drought, salinity, heat, cold, etc.) stresses account for about one-fourth of the value of agricultural produce. The storage, transportation, processing, value addition and marketing of farm produce need to be improved to enhance household food, nutrition and livelihood security. Indian agriculture is characterized by agro-ecological diversities in soil, rainfall, temperature, and cropping system. Besides favorable solar energy, the country receives about 3 trillion m^3 of rainwater, 14 major, 44 medium and 55 minor rivers share about 83 per cent of the drainage basin. About 210 billion m^3 water is estimated to be available as ground water. Irrigation water is becoming a scarce commodity and no suitable watershed technology developed so far. Thus proper harvesting and efficient utilization of water is of great importance in near future.

Intensive cultivation introduce of high yielding varieties in the mid 1960's required higher energy inputs and better management practices. Land preparation, harvesting, threshing and irrigation are the operations, which utilize most of the energy used in agriculture. For desired cropping intensity with timeliness in field operations, animate energy sources alone were no longer adequate. Farmers opted for mechanical power sources to supplement animal power. Small and marginal farmers have limited resources especially in rain-fed regions where only animal power is used resulting in low productivity. Though agricultural production is high, per hectare productivity is much lower than world average due to several reasons. There is an urgent need to increase productivity with changing advance technology. Smaller the farm, greater is the need for marketable surplus, so that small farmers can have a reasonable income to achieve this goal will be possible only if we develop and disseminate eco-technologies rooted in the principles of ecology, economics, gender equity and employment generation. The estimated food requirement in India and total production of major crops indicate that to keep pace with the present population growth and consumption pattern, food grain requirement has been estimated to be 240 MT by 2020 and 300 MT by 2025. Annual agricultural growth should be maintained at 6.7 per cent to meet these demand projections for future. Non-availability of manpower during peak crop season is many times a problem and urgent need of fully mechanization.

The overall achievement in the creation of irrigation facilities has been relatively better in India with 63 per cent growth rate compared to the world average. There is a need to increase the utilization of rainwater to enhance the gross cropped area by 30 M ha as the yield of food grain in irrigated areas is almost twice that in rain-fed agriculture. The infrastructure for agricultural diversification, reducing post harvest losses of perishables, value addition to agro- produce and branding system needs to be strengthened with a support of modern technologies.

(i) Farm Mechanization Available

Mechanization is viewed as package of technology to ensure timely field operations to increase productivity and reduce crop losses and improve quality of agro- produce. Increase land utilization and input use efficiency using farm mechanization. Increase labour productivity using labour saving and drudgery reducing devices besides, being cost effective and eco-friendly. Appropriate machinery have been adapted by farmers for ensuring timely field operations and effective application of various crop production inputs utilizing human, animal and mechanical power sources. Diverse farm mechanization scenario prevails in the country due to size of farm holdings and socio-economic disparities. Indian agriculture continues to be dependent upon farm labor and draught animal power. Hand tools and animal drawn implements are extensively used which involve a great amount of drudgery. Adoption of tractors has been on the increase throughout the country. The availability of farm power or energy per unit area (kW/ha) has been considered as one of the parameters for expressing level of mechanization. Ratio of machine work and sum of manual and machine work has also been considered to estimate the level of mechanization. The ratio of tractor power to total farm power has increased from 0.82 to 32.25 per cent. A number of machines and equipment suitable for different farm operations suited to different agro-climatic regions and categories of farm have been developed. Most of these machines are available commercially and have been well adopted according to regions. Bull dozers, power rakes, land planes, laser leveler, harvester, planter, and reaper etc. are commercially available with different capacities. India has made remarkable progress in agricultural mechanization technology. One of the globally used Index of Agriculture Mechanization (IAM) is power availability per unit area. The power availability in hilly areas is also quite low due to undulated land. The increasing use of tractors and irrigation pumps operated by electric motors and diesel engines are the indictors of the fact that use of mechanical power in India has increased many fold during the last two decades.

(ii) Irrigation Facility Available

Water is an essential natural resource for the plant as well as animal. Water is a renewable sources, it is quite dynamic and scarce on earth. The source of all water is annual rainfall and rainfall in India is highly variable, irregular and undependable with widespread variation among various meteorological sub-divisions. The highest and lowest annual average rainfall in India is 10,000 mm (Khasi-Jaintia Hill, Meghalaya) and 100 mm in Rajasthan. The distribution of water to make it better the technical feasibility of inters basin transfer of water by linking Himalayan and Peninsular rivers. Every drop of rainfall should be conserved and it can be done by *in-situ* and *ex-situ* harvesting of rainfall and significant progress has been made in the development of water harvesting, conservation and utilization technology in agriculture sector. For higher water use efficiency, it is necessary to adapt efficient irrigation methods that are technically feasible, economically viable and socially acceptable. Drip and micro irrigation systems need to be adopted and popularized for row crops, horticulture and especially for widely spaced high value crops in undulating terrains, shallow and porous soils and in water scarce areas.

Floods normally affect 8 major river valleys spread over 40 million hectares of area in the entire country affecting nearly 260 million people. Similarly, drought affects 86 million people who are spread in 14 states covering a total of 116 districts. If we have to prevent the damage due to the flood and reduce the severity of drought, we have to harness this 1500 bcm of water and distribute it to the drought affected areas. Therefore, National Water Grid (NWG) needs to be developed by interlinking of the country's major river basins. In India, major water resources are from rivers, lakes, canal, reservoirs, tanks and ground water. Globally, fresh water at a tune of 3,240 m km^3 is being utilized. Of this, 69 % is being used in agriculture sector, 8 % in domestic, 23 % in industrial and other sectors. In India, around 88 % water is being used in agriculture sector, covering around 80 M ha area under irrigation. Due to liberalization of industrial policies and other developmental activities, the demand for water in industrial and domestic sectors is increasing day by day, which forces the reduction of the percentage area under irrigation in the agriculture sector. Thus, pressure lies on agriculture sector to reduce the share of water and increase the production, which could be achieved by enhancing increased water use efficiency.

Water is drawn either from a dug well or a shallow tube well fitted with a pumpset. In both cases a horizontal centrifugal pumpset is commonly used. It was estimated that there are about 16.0 million electric motors and 9.0 million diesel engine pump sets for lifting water from various sources as of 2003–04. These consume about 100 billion kWh of electricity and 4.0 billion liters of diesel annually. Several field studies revealed that most of the agricultural pump sets selected and installed operated at much lower efficiency than achievable. The standing committee (Government of India) on the study of operational efficiency of irrigation pumps clearly observed that overall efficiency in diesel operated pump sets was 12.7 % and 31.1 % in electrically operated pump sets. The efficiency can be increased by selection of appropriate pump, proper installation of system, selection of prime mover, selection of proper couplings, and selection of proper size of suction and delivery pipes and regular maintenance. Due to improper management of the systems, lack of clear cut guidelines and lack of emphasis on water management at farm level, the overall efficiency of these systems continued to remain low. It is reported that, a farmer can save up to Rs. 3300 per year over his present expenses for pumpset operation if his pumping system is properly selected. Such correct selection would not only salvage the farmer from his avoidable yearly financial loss but would also save the nation a few hundred cores worth of power by way of savings in diesel oil consumption and electric power.

(iii) Energy in Agriculture

Fossil fuels also covers electricity generation are extensively used, but these resources are finite. It is estimated that 81792 million units of electricity and 40.12 million tones of high-speed diesel are available in the country to meet the energy demand in 2005-06. The studies undertaken on Energy Requirement in Agriculture has indicated that use of energy for crop production varied considerably from 9600 MJ/ha to as high as 21000 MJ/ha including seeds and fertilizer. Energy in fertilizer contributed the maximum followed by irrigation. The average operational energy use on all India basis for crop production increased from 3,374 to 8,138 MJ/ha during 1970 to 1996 period. More than 600 million tones of biomass are available from various crops residues and agro-wastes for fuel wood and animal feed. It is estimated that 35-40 per cent of bio- mass is utilized for animal feed and the remaining as energy source through direct combustion either for cooking food improved chulha 284.89 lakhs or for processing of agro-produce bagasse co-generation 84 MW. This is also available for generation of 27.13 lakhs plants bio-gas and gasification 28.53 MW. Bio-gas and producer gas can partially substitute HSD in compression ignition engines. Alcohol and plant oils are compatible with fuel injection system of compression engines and can be used after transesterification. Solar dryers and solar cookers 456674 numbers are available to supplement rural energy.

(iv) Rural Energy Available

Energy in the rural sector is required for rural home management, production agriculture, cottage industries and agro-processing. It is estimated that about 66-80 per cent of the total energy for the rural sector is used mainly for rural home management and 16-25 per cent for agriculture production. These are met by bio- energy, natural energy, electricity, fossil fuel and coal. Cooking alone consumes 70 per cent of the total energy, which is met by wood, crop residues, and dry animal dung. About 3-5 kg of fuel wood per household is required for every day. Yearly requirement of fuel wood for rural sector alone will exceed 200 million tones. The rural energy needs are different from urban needs due to economic disparity. As per NSSO Report No 404, 1997, about 78 per cent of rural households use fire wood compared to 30 per cent in urban areas for cooking and about 31 per cent rural homes use electricity for lighting compared to 85 per cent in urban areas. Only about 86 per cent villages are electrified. The remaining households use kerosene. This requires about 4 liters of kerosene per month per family. This would require more than 300 million liters of kerosene in rural homes for lighting. One of the reasons of poor industrialization of rural sector is due to inadequate or non- availability of assured electricity in the villages.

(v) Agro-processing Industry

The Indian food agro-products industry is predominantly related to conservation and simple processed food. However, there is a trend towards value added, easy to use convenient products. Some of the major post-harvest technology and equipment developed in India are: Optimum stage of harvesting, Status of traditional post harvest activities, Effect of chemicals on yield, Groundnut stripper, Maize dehusker sheller, Crop covering structures, Storage Structures, Crop driers, Cleaners and graders, Decorticators and threshers, Storage structures for onion, Solar drier and seed treater, Milling equipment for small scale operations, Leaf grinder and pyrolyser, Decorticators, Grain infestation detector, Straw baler and feed treatment structure, Pedal-cum-power operated cleaner, seed treater, Use of biogas in grain storage, Bottling of sugarcane juice, Rice puffing machine, Pantnagar dal mill, Pearler for coarse grains, Sago roaster, Ginger and turmeric polisher, PKV dal mill, Pea peeling and punching, Extraction of chillies seed, Agro Processing Centre, Garlic bulb breaker, use of waste Kagzi lime for citric acid, Enhancement of shelf life of fruits and vegetables, Cashewnut sheller, Water chestnut decorticator, Mango grader, Bengal gram stripping-shelling machine, magnetic treater for seeder, Microwave for disinfestation, Honey bee smoker, Radial honey extractor, Effluent treatment system, Blended leather boards etc.

Post harvest characteristics of agricultural produce is affected by pre-harvest treatments-seed rates, level of fertilizer use, nutritional balance, irrigation and drainage, attack f diseases and pests, growth hormones and pesticides used an d their residual toxicity, mechanical and environmental injuries during harvesting, handling, transport and storage. A strong awareness drive is needed to provide pre-harvest inputs for scientific post-harvest management of agricultural produce and by- products.

For scientific storage of agricultural produce cleaning, grading, shelling, decortications, drying to safe moisture levels -10-12 per cent in case of cereals, 8-10 per cent in case of pulses and 6-8 per cent in case of oilseeds, are important. A number of useful equipments-for such unit operations are available that need to be commercialized and extended to targeted beneficiaries. Scientific storage-room, warehouses storage bins for growers to negotiate with the forces of market enabling them to earn 25-50 per cent more net returns.

The utilization pattern of produce and their by-products governs the use of process and machinery. By improved harvesting, parboiling, drying and milling technologies the total out-turn of rice can be increased by 10 per cent. There is an increasing trend towards the use of roller mills for milling wheat that give quality products. Maize is a very versatile grain having food, feed, and industrial raw material value. It yields numerous products employing dry milling, wet milling, fermentation and other processes. Sorghum and minor millets have excessive fibre causing anti-nutritional effects; to overcome this defect peelers are used.

(vi) Priority in Post Harvest Technology Research and Development

Post harvest technology (PHT) is commodity and location specific for availability of raw materials and higher profit. Development of need-based and market driven post harvest technology and equipment for loss prevention and value addition to raw food materials of plant, animal and aquatic origin for internal and international markets. Diversification in the present uses of rainfed and other crops may be considered. The technology so developed must lead to rural industrialization, thereby creating employment and income generation opportunities. Appropriate PHT would help in enhancing per capita food and fibre availability from the limited and dwindling land and water resources.

(vii) Food Processing Industry

Adoption of innovative and scientific packaging methods by food industry has enabled the manufacture of safe and quality food for value addition. More than 73 per cent of rice, 55 per cent maize, 24 per cent pulses, 45 per cent oil seeds, 45 per cent sugarcane are processed by modern machinery besides other commercial crops. The large food industry has come up in a broader way with meeting international quality standards, and exporting processed foods even to the developed world. The food industry can be considered to be bright only if safe and quality food products are made available covering the complete chain starting from raw material production to delivery of safe food and poison free to the end user. Indian groundnut processors are able to meet the restriction of 5 ppb afflatoxin level in peanut butter and other byproduct of groundnut. Now, our processors have adopted a vacuum packaging approach for roasted and fried groundnut kernel based snack products to check off-flavor development. It should not be a matter of surprise if Indian, paddy processing facilities of International Standard is widely found. High value Basmati paddy is milled and exported with the quality standards prescribed by different International Markets like USA, Middle East, etc. The food processing industry in India is labour intensive and offers a major employment opportunity. The industry employs 18 to 20 per cent of country's labour force. It is estimated that per Rs. 1000 Core investment in the large-scale food industry provides direct employment to 54000 persons whereas in cottage scale the employment potential is 354000. To provide food security through financial empowerment the Govt. is keen to provide employment many through this sector. Also, normally, producers do not get enough income /returns due to low price or glut. Under such circumstances, it is a must to initiate processing activities in the production catchments to improve the raw material quality and enhance shelf life. Also, these units can take-up the activities of supplying primarily processed good quality raw material to large-scale industries.

(viii) Opportunities of Employment and Income Generation

The agriculture processing sector has huge employment potential for rural people, provided the primary processing activities are undertaken, as in the past, in rural areas. Cottage and industrial level primary and secondary processing include, rice mills, grain mills, fruit and vegetable processing, etc. The case of employment generation by food grain processing in production catchments was worked out, showing huge potential even for processing 20-50 per cent of total food grain production in rural area. However, a systematic approach needs to be considered. The low price pattern in the production season, grape growers in Maharashtra State have taken-up on-farm primary processing activities for producing raisins. Quality aspects are kept in mind all throughout and the end product processed in rural areas can be exported. It is important to appreciate that 'growing' and 'processing' are complementary and need to be promoted together for more income to the poor peoples.

Primary processing of cereals, pulses, oil seeds, fruits and vegetables can be conveniently taken up for value addition and income generation.

(ix) Women Friendly Improved Tools and Equipment for Agriculture

Women play a major role in rural India through their active participation in agriculture. At present, the women found in the work force in agriculture and allied sectors are estimated at about 61 million which amounts to about 30 per cent of the total rural workers in the country. They participate in different crop production and food processing operations as well as in animal husbandry and dairy and fishery activities. Studies have shown that the Indian women work for about 14-16 hours a day to carry out various activities on farm and at home. Rural women are usually employed in field operations like sowing behind the plough (Pora and kera), transplanting (paddy, vegetables and fruits), weeding, interculture (thinning, roughing), harvesting and threshing. The activities in agro processing involve cleaning, grading, drying, parboiling, milling, grinding, decortications and storage. Women workers are also preferred in commercial agriculture like tea, coffee, tobacco, orchard and plantation crops. Bidi making, jute retting, lac cultivation, processing and lac products preparation, cotton picking, sugarcane detopping and spice picking, with cleaning and processing also largely performed by women with expertise. There are more than 50 improved hand tools and equipment developed by various research organizations in the country out of these, 30 hand tools have been identified which can be made suitable for women workers.

CHAPTER 8

Crop Significance and Classifications

Agricultural crops are very essential to know the significance in our life. In India, slow agricultural growth is a major concern for policymakers as two-thirds of India's people depend on rural employment for a living. Agricultural practices are neither economically nor environmentally sustainable. Poorly maintained irrigation systems or we can say fully depend on monsoon rain is the main factor responsible for its slow growth. Crops which are grown in different parts of India, contributes to overall growth of the country's economy. Since prehistoric times, Indian economy was often called as 'monsoon economy'. In India irrigation systems are very poor and water use efficiency is less. Agriculture is fully depended on monsoon rain which is the major factor responsible for its slow growth and sometime consequent to drought situations in some parts of the country. The livelihood of a vast population in India depends on agriculture, forestry, wetlands and fisheries and land use in these areas is strongly influenced by water-based ecosystems that depend on monsoon rains.

Food crop like rice is the main grain crop of India. India ranks second in the world in production of rice. About 34% of the total cultivated area of the nation is under rice cultivation. Rice is cultivated in areas having annual average rainfall of 125 cm. Major rice cultivating areas are north east India, eastern and western coastal regions. West Bengal, Punjab and Uttar Pradesh are the major rice producing states. Wheat is the second major crop in India. Wheat is cultivated in areas with mean annual rainfall of 75 cm and fertile soil. Wheat has got an important role in 'Green Revolution'. The highest quantity of wheat in the country is in Uttar Pradesh. 35 % of wheat is produced only in Uttar Pradesh. This is produced by Punjab and Haryana where production of wheat is on a large scale. Maize is an important khaki crop of rainy season. Maize is cultivated in different areas and in different climates but it is suitable where temperature is 35° Celsius and rainfall is 75 centimeters. It is cultivated in hilly areas-of Jammu and Kashmir and Himachal Pradesh. Maize is cultivated throughout our country but it is cultivated more in Punjab, U.P., Bihar, M.P. and Rajasthan. Pulses are grown in dry climate region in India. These crops provide nitrogen to the soil. Pulses are a source of proteins in the diet. Madhya Pradesh is the leading pulses producing state in India. It is also produce in Uttar Pradesh and Rajasthan. Jower is grown where the climate is hot and dry. It is cultivated in Maharashtra, Karnataka, Andhra Pradesh, and Tamil Nadu.

In agriculture there are few crops which are grown for profit are called as 'Cash Crops or Commercial crops'. Cash crop is a backbone of agriculture economy of India. It sets a strong base for Indian economy where country's trade and commerce flourish domestically and internationally. Cash crops are generally grown for money. In earlier days, cash crops were grown in a very small scale but today it forms a major contribution to our nation's economy. Now it has grown at large scale for commercial purpose.

Cash crop like sugarcane is an important cash crop of India. Molasses, sugar and khandasari etc. are produced from the juice of sugarcane. Sugarcane cultivation needs temperature of 15° to 40° and rainfall of 100 to 150 centimeters and fertile loamy soil or hard soil. Sugarcane is cultivated from Kanyakumuri to Punjab but it is more cultivated in Uttar Pradesh. Coffee has a great demand in the world market. For reason, India exports coffee. Coffee cultivation needs hot and wet climate and fertile sloppy land. It is mainly produced in south Indian states of Karnataka, Kerala and Tamil Nadu. Coffee is produced on a large scale on the mountain ranges of Nilgiri. India is first in the cultivation of tea in the world. Tea cultivation needs hot climate, excess rainfall and sloppy soil. Tea is found more in Assam, Karnataka, Kerala and Himachal Pradesh, Dehradun, Ranchi of Bihar and Tripura. Rubber is needed by different industries and transport industry in this modern age. It is cultivated in hot and wet climate garden in natural way. It is cultivated in the State of Kerala in India. Except Kerala, Andaman Nicobar Islands, Kurgan of Karnataka State and Chicmagalur district etc. Oilseeds crops like groundnut, mustard, rapeseed, linseed and caster help us to get our edible oil. Oil is also extracted from coconut. India occupies the first position in the world in the production of groundnut. Groundnut cultivation needs temperature varying from 20° to 30° degree Celsius and needs 60 to 80 centimeters of rainfall. Groundnut cultivation needs sandy and light soil. Groundnuts are produced more in Tamil Nadu, Maharashtra, Gujarat, Andhra Pradesh and Uttar Pradesh.

Fibre crop like cotton needs 20° to 25° Celsius temperature and 50 to 75 centimeters of rainfall. Cotton plant needs wet climate at the time of growing and dry climate at the time of collecting seeds. it is known as *Kharif* crop. Cotton cultivation takes place in Gujarat, Maharashtra, Punjab and Haryana of our country. Except these States, cotton cultivation also takes place in Karnataka, Tamilnadu, M.P, Rajasthan, A.P. and U.P. India is a major producer of jute in the world. It is another type of fiber crop. Bags, ropes and a lot of other things are made out of jute. Jute cultivation needs hot and wet climate and fertile loamy land. Temperature of 24° to 35° Celsius and 90 to 150 centimeters of rainfall is suitable for jute cultivation. In our country, jute is cultivated in Orissa, West Bengal, eastern U.P., Bihar, Assam and Tripura.

(1) Crop Significance in Human Diet

Human nutrition is most important to our daily life. Many agricultural crops and their products find place in our meals and diet. Human body requires vitamins, minerals, proteins, energy etc. for its health. All these are supplied by agricultural crops, horticultural crops fruits and vegetables are the chief sources of vitamins, minerals, carbohydrates, fats, proteins etc. are recognized as protective foods as they are necessary for the maintenance of human health. Cereal crops supply carbohydrates, protein etc, oilseed crops supply fat and pulse crops supply protein etc. A person should consume at least 120 g of fruits per day as per Indian Council of Medical Research (ICMR), but the consumption of fruits per capita in India is still low. However, the availability of fruit is 172 g per day in our country. Deficiency of any minerals and nutrients is depicted by the human body by giving typical symptoms. The great majority of people obtain most of their carbohydrates and proteins from cereals and pulses but their diets must also contain significant amount of fruits to ensure that they get the vitamins which are not provided by the staple cereal foods.

Vitamins

These are the important constituents of fruits and vegetables and are indispensable part of human diet. Although required in very minute quantities, they are absolutely essential for the maintenance of health. The deficiency of any vitamin from the diet for considerable period may lead to diseased state or disorder conditions. Fruits and vegetables supply several vitamins.

Vitamin-A

It is essential for normal growth, reproduction and maintenance of health and vigor. It affords protection against cold and influenza and prevents night blindness. Its deficiency results in– cessation of growth in young children, night blindness, dry up of tear glands in the eyes, eruption of skin and brittleness of the teeth. The sources of vitamin-A is mango, papaya, dates, jackfruit, walnut etc.

Vitamin B1 (Thiamine)

Tones the nervous system and helps in proper functioning of the digestive tract. Its deficiency in human diet results in Beri-beri, paralysis, loss of sensitivity of skin, enlargement of heart, loss of appetite, loss of weight and fall in body temperature. Source of this vitamin is orange, pineapple, jack fruit, cashew nut, walnut, dry apricot, almond, banana etc,.

Vitamin B_2 (Riboflavin)

Vitamin B_2 is required for body growth and health of the skin. Its deficiency causes sore throat, anorexia cataract, and loss of appetite and body weight and also development of swollen nose. Source of this vitamin is bael, papaya, litchi, banana, apricot, pomegranate, pear etc.

Vitamin B6 (Pyridoxin)

Its deficiency causes dermatitis, anemia, ulceration in oral cavity etc. Vitamin rich fruits are chestnut, walnut, almond, apricot, apple, plum etc.

Vitamin B_6 (Pyridoxin)

Its deficiency causes dermatitis, anemia, ulceration in oral cavity etc. Vitamin B_6 rich fruits are chestnut, walnut, almond, apricot, apple, and plum.

Vitamin B_3 (Niacin)

Deficiency symptoms of this vitamin are sour tongue, Pellagra, complex of diarrhea, loss of mental aptitude, dermatitis, discolouration of skin of hands, feet and legs and under severe condition the mental balance may shift. Source of this vitamin rich fruits are chestnut, walnut, almond, apricot, apple, and plum.

Vitamin B_9 (Folate or folicin or folic acid)

Source of this vitamin is dark-green leafy vegetables such as spinach, mustard greens, butter head lettuce, broccoli, Brussels sprouts, and okra, legumes such as cooked dry beans, lentils, chickpeas and green peas. Deficiency symptoms of this vitamin is birth defects, cancer heart disease, and nervous system

Vitamin -C (Ascorbic Acid)

This vitamin promotes general health and healthy gums, Prevents scurvy disease which is characterized by pain in the joints and swelling of limbs (rheumatism), bleeding of gums, tooth decay and keeps the blood vessels in good condition. Source of this vitamin is amla, guava, ber, citrus, strawberry, and pineapple.

Vitamin K

Vitamin K plays an important role in helping blood the clot and prevents excessive bleeding. Source of this vitamin is nuts, meats, cheeses, and eggs and its constituent synthesis of pro-coagulant factors. Low intake of Vitamin K results osteoporosis and induced bone loss.

Vitamin E (tocopherols)

Source of this vitamin is almonds, cashew nuts, filberts, macadamias, pecans, pistachios, and walnuts heart-disease, LDL oxidation, immune-system, diabetes, cancer

Fibre

Source of this fiber is most fresh fruits and vegetables, nuts, cooked dry beans and peas.

Mineral Nutrition

A mineral is a chemical element required as an essential nutrient to perform function necessary for life. There are five major nutrients essential for human body is calcium, phosphorus, potassium, sodium, and magnesium. Among five major nutrients three have been discussed.

(i) Calcium

Source of calcium is papaya, raisins, orange and almonds. Symptoms are osteoporosis, muscular/ skeletal, teeth, and blood pressure.

(ii) Magnesium

Source of this magnesium are banana, nuts, cashews. Major symptoms are osteoporosis, nervous system, teeth, and immune system

(iii) Potassium

Source of this mineral are dried fruits (such as apricots and prunes). Major symptoms are hypertension (blood pressure) stroke arteriosclerosis.

(2) Crop Significance in Economic Importance.

It is possible to increase production by growing more number of crops on same piece of land or by vertical expansion. With the support of agriculture, many agroindustries can spring up in villages itself. It is an acceptable fact that agriculture can come in a big way to solve the problem of unemployment. Source of other industries e.g. rubber, oil, gum, dyes, chemicals etc. raw material for fruit and vegetable processing plants. Employment is also generated in field operations like fruit picking, harvesting, grading, packing and selling etc.

Dependence on cereals can be decreased since the fruits and vegetables provide almost every nutrient and mineral required for the human body as the food. Foreign exchange can be earned by exporting the fruits and vegetables and their quality products e.g. mango, almond, saffron etc. Products like jam, jelly, pickles etc. are exported and similarly making them indigenously thus saving the foreign exchange minimizes the import of wines and other agriculture-based products.

Areas related to agriculture for providing employment in Agronomic crop cultivation, Fruit cultivation, Vegetable cultivation, Plantation and spice crop cultivation, Vegetable F1 seed production, Vegetable and flower seed production for export, cut flower production and floriculture, Landscaping, Nursery raising, Tissue culture, Marketing, Processing of agriculture produce, Government development departments, Ancillary services i.e. fertilizers, tools, insecticides, pesticides , irrigation equipments, chemicals like growth regulators etc. manufacture , sale and consultancy services, and Research and education.

(3) Crop Significance in Recreation

Roaming in the gardens, orchards or places well planted with flowerbeds etc. gives mental piece to the persons. One enjoys fresh air and natural beauty, sheds of tension making him fresh.

(4) Crop Significance in Medicines

The parts like stem, leaf, flowers, roots and even the fruits of agriculture plants are used to make drugs, chemicals, insecticides, germicides etc. e.g. rose water is used to cure eyes ailments. Similarly saffron is important ingredient of many medicines. Papain is a digestive enzyme, citrus fruit like sweet lime is used for liver ailment, rind of pomegranate and pectin from guava used for stomach upset, bark of arjun trees for heart troubles, neem water for skin irritation and allergies etc.

(5) Crop Significance in Aesthetic Value and Religious Importance

Aesthetic value and religious importance is the unique factor distinguishing it from agricultural activities. Mango leaves, wood, banana leaves etc. are used for religious functions. Similarly the plantation of banana tree in the court yard or tulsi plantation is said to bring prosperity as per Hindu religion. This aspect of agriculture has lead to its universal popularity. Paradise means garden. The hanging garden of Babylon (one of the 7th wonders of the world) and the Versailles garden of 17th century are unique examples.

Classification of Crops

The crops are classified based on season, taxonomy, botany, special purpose, cultural method/water, root system and cotyledon.

(1) Classification Based on Season

(a) **Kharif Crops:** The Kharif/monsoon crop is sown with the beginning of the first rains in June to July, during the south-west monsoon season and harvest in the month of October. The term Kharif means "autumn" in Arabic, e.g. Paddy, Maize, Soybean, Sorghum, Groundnut, Cotton and Urd etc.

(b) **Rabi Crops:** The Rabi/winter crop or spring harvest crop is sown in the month of November during the north-east monsoon season and harvest in the month of February. The term Rabi means "spring" in Arabic, e.g. Wheat, Chickpea, Pea, Sunflower, Safflower, Mustard etc.

(c) **Zaid Crops:** The Zaid/summer crop is mainly sown in the month of March and harvested in May. Requires warm day weather for major growth period and longer day length for flowering e.g. Paddy, Groundnut, Moong, Sunflower, Maize, Groundnuts, Watermelon, Pumpkins and Gourds.

(2) Classification According to Taxonomy

(a) Asteraceae/Composite : Sunflower, Safflower, Niger

(b) Cruciferae :Mustard, Radish, Cabbage, Cauliflower etc.

(c) Cucurbitaceous : bottle gourd, Bitter ground, Pumpkin etc.

(d) Chenopodiaceae : Sugar beet, Beet, Spinach

(e) Euphorbiaceae : Castor, Topioca

(f) Malvaceae : Cotton, Ladyfinger, Rosette

(g) Papilionaceae/ Leguminoceae : Pea, Gram, Arhar, Groundnut, Berseem, Lathyrus, Sunhemp, Lucerne, Urd, Moong, Lentil, Soybean etc.

(h) Linaceae : Linseed

(i) Pedaliaceae : Sesamum (Til)

(j) Poaceae/Graminae : Cereals, Millets. Grasses, Sugarcane, Napier, Oat

(i) Polygonaceae : Buckwheat

(l) Solanaceae : Potato, Tobacco, Tomato, Chili, Brinjal

(3) Classification According to Botany

(a) Cereal Crops:

(i) Rice : Oryza sativa L.

(ii) Wheat : Triticum aestivum L.

(iii) Maize : Zea mays L.

(iv) Bajra/Pearl millet : Pennisetum typhoid's/P. glaucum L.

(v) Sorghum/Jowar : Sorghum bicolor/S. vulgare L. Moench

(vi) Barley : Hordeum vulgare L.

(vii) Triticale : Secale cereal

(viii) Buckwheat/Pseudo cereal : Fagopyrum esculentum

(b) Millet Crops:

(i) Cheena/Proso millet : Panicum miliacearum

(ii) Foxtail /Italian / Jerman millet/Kakun : Seteria italica L. Beauv

(iii) Kodo/Coarsest millet : Paspulum scrobiculatum L.

(iv) Little millet : Panicum sumatrense

(v) Madua / Ragi / Finger millet : Eleusine coracana Gaertn

(vi) Sawan/Barnyard millet : Echinochloafrumentance L.

(c) Pulse Crops:

(i) Gram/Chickpea/Bengal gram : Cicer aeritinum L

(ii) Field Pea/Grain pea : Pisum sativum var. arvense

(iii) Arhar/Pigeonpea/Red gram : Cajanus cajan L. Millsp.

(iv) Soybean : Glycine max L. Merril

(v) Black gram/Urdbean : Vigna mungo/Phaseolus mungo L. Hepper

(vi) Green gram/Moong/Moongbean : Vigna radiatefPhaseolus aureus L. Wilczek

(vii) French bean/Rajmash : Phaseolus vulgaris

(viii) Indian Cowpea/Lobia : Vigna unguiculata/V. sinensis L.

(ix) Lentil : Lens esculantum/L. culinaris Moench

(x) Lathyru s / Chickling pea/Grasspea : Lathyrus sativus

(xi) Mothbean : Vigna/Phaseolus aconotifolia

(xii) Horse gram/Kulthi : Macrotyloma uniflorum

(d) Edible Oilseed Crops:

(i) Groundnut/Peanut/Monkey-nut : Arachis hypogea L

(ii) Sunflower : Helianthus annus L

(iii) Safflower : Carthamus tinctorius L

(iv) Rapeseed and Mustard : Brassica spp. L.

(v) Sesamum/Til : Sesamum indicum L.

(vi) Niger : Guzotta abssicinia

(vii) Linseed/Flex : Linnum ussitatisimum L

(viii) Non-edible Oilseed Crops

(ix) Castor : Ricinus communis L.

(e) Fiber Crops:

(i) Cotton : Gossipium spp.

(ii) Jute/tita pat : Corchorus capsularis

(iii) Sunhemp: Crotolaria juncea L.

(f) Forage Crops:

(i) Berseem : Trifolium alexandrinum L.

(ii) Lucerne/Alfalfa : Medicago sativa L

(iii) Oat : Avena sativa L.

(iv) Napier grass : Pennisetum purpureum L

(v) lusterbean/Gaur : Cymopsis tetragonalaba L.

(g) Sugar Crops:

(i) Sugarcane /Cane : Saccharum officinarum L.

(ii) Sugar-beet : Beta vulgaris L.

(h) Tuber Crops:

(i) Potato : Solanum tuberosum L.

(ii) Tapioca : Manihot utilissima

(i) Stimulate Crops:

(i) Tobacco : Nicotiana spp.

(ii) Opium : Papaver somniferum

(j) Medicinal Crops:

(i) Safed musli : Chlorophytum borivilianum

(ii) Ashwagandha/Winter cherry : Withania somnifera

(iii) Rouvolfia / Sarpagandha : Rouvolfia serpentine

(iv) Isabgol : Plantago ovate

(v) Butch : Acorus calamus

(vi) Bramhi : Bacopa morriei

(vii) Nux vomica : Strychnos nuxvomica

(k) Aromatic Crops:

(i) Lemon grass : Cymbopogan flexuasus

(ii) Mentha/Mint : Menthe arvensis

(iii) Khus/Vetivar : Vetiveria zizanoides

(iv) Citronella : Cymbopogan ivinterianus

(v) Tulsi/ Basil : Ocimum sanctum

(4) Classification Based on Special Purpose

(i) Arable Crops: Crops which are cultivated on ploughed land. They are annual crops and include cereals, root crops, tobacco, sugarcane, maize and potatoes.

(ii) Aromatic Crops: The crop/plants contain odoriferous and volatile substances, which occur as essential oils, gum exudates; balsam and oleoresin in one or more part of plant, *viz.* wood, bark, foliage, flower and fruit are called Aromatic plants.

(iii) Alley Crops: Alley crops or hedge-row intercrops is grown under an agro forestry practice in which perennial, preferably leguminous trees or shrubs are grown simultaneously with arable crop. The trees, managed as hedgerows, are grown in wide rows and the crop is planted in the interspaced or 'alley' between the tree rows. e.g., Sweet potato, Black gram Turmeric and Ginger are grown in the passages formed by the rows of Eucalyptus, Subaru and Cassia etc.

(iv) Augment Crops: Such crops are grown to supplement the yield of the main crops, e.g. Japanese mustard with berseem.

(v) Avenue Crops: Such crops are grown along farm roads and fences *e.g.*, Pigeon pea, *Gliricidia sisal* etc.

(vi) Border/Guard Crops: Such crops protect another crop from trespassing of animals or restrict the speed of wind and are mainly grown as border *e.g.*, safflower (thorny oilseed crop) is planted around the field of gram.

(vii) Cash Crops: A crop, such as tobacco, grown for direct sale rather than for livestock feed or a crop grown by a farmer primarily for sale to others rather than for his or her own use *e.g.*, sugarcane, cotton, jute, tobacco etc.

(viii) Catch/Contingent Crops: Such crops are cultivated to catch the forthcoming season when main crop is failed *e.g.*, Linseed, toria, urd, moong, cowpea etc.

(ix) Contour Crops: Crops are grown on or along the contour line to protect the land from soil erosion *e.g.*, marvels grass etc.

(x) Cover Crops: A close-growing crop grown primarily to improve and protect the soil from erosion through their ground covering foliage and/or root mats between periods of regular crop production *e.g.*, Labia, groundnut, urd, sweet potato, methyl etc.

(xi) Complementary Crops: Both main and intercrop is benefited to each other *e.g.*, Jowar + Labia.

(xii) **Competitive Crops:** Such crops compete to each other and are unsuitable for intercropping *e.g.*, two cereals.

(xiii) **Exhaustive Crops:** These crops leave the field exhaustive after growing *e.g.* Cereals, (Rice).

(xiv) **Energy Crops:** An energy crop is a plant grown as a low cost and low maintenance harvest used to make befouls, or directly exploited for its energy content *e.g.*, sugarcane, potato, maize, topioca.

(xv) **Fouling Crops:** Such crops whose culture practices allow the infestation of weeds intensively *e.g.*, direct seeded upland rice.

(xvi) **Ley Crops:** Any crop or combination of crops is grown for grazing or harvesting for immediate or future feeding to livestock *e.g.*, Berseem + Mustard.

(xvii) **Medicinal Crops:** The crop/plant contains alkaloids, glycosides, steroids or other groups of compounds of medicinal value, which is used commercially, such plants are called Medicinal plants.

(xviii) **Mulch Crops:** Such crops are grown to conserve the soil moisture through their ground covering foliage *e.g.*, cowpea.

(xix) **Nurse Crops:** A crop of trees (nurse trees) shrubs or other plants introduced to foster or nourishment of another crops by i.e. shading it protecting it from frost, insulation or wind. The widest use of nurse corps is in the establishment of leguminaceous plants such as alfalfa, clover *e.g.*, Sun hemp in sugarcane, Jowar in cowpea, Rai in pea.

(xx) **Paira/Uteri Crops:** The seed of succeeding crops like lentil, gram, pea, lathy us, bireme, linseed etc. is sown broadcast at 10 to 15 days before harvesting rice crop. This practice saves time; money (to be spent on land preparation etc.) utilizes residual fertility. This practice is common in both upland and lowland rice culture.

(xxi) **Paired row Crops:** Generally the third row of crop is removed or growing of crop in pair row and the third row is escaped with an object to conserve the soil moisture in Dry land areas.

(xxii) **Restorative Crops:** Restorative crops are crops that help in maintaining the fertility of the soil, for *e.g.*, pulses and legumes.

(xxiii) **Silage Crops:** Such crops like corn, legumes, and grasses that have been harvested at early maturity, finely chopped, packed tightly to exclude air, and stored in tower silos, pits, or trenches for properly fermentation which is used as animal feed during lean period or off season *e.g.*, Maize, cowpea, Jowar etc.

(xxiv) **Smother Crops:** Smother crops are specialized cover crops being ability to suppress weeds by providing dense foliage and quick growing ability *e.g.*, buckwheat, mustard, cowpea, urd etc.

(xxv) **Stimulate Crops:** Stimulate crops stimulate the human body *e.g.*, tobacco, opium etc.

(xxvi) **Supplementary Crops:** Such cops are neither complementary nor competitive *e.g.*, maize + cucurbits.

(xxvii) **Trap Crops:** Trap crops are grown to protect the main cash crop from a certain pest or several pests. These crops are planted in completely surrounding the main cash crop and prevent pest attack from all sides of the field through attracting the pest *e.g.*, cotton red bug trapped by Ladyfinger around cotton.

(xxviii) **Truck Crops:** Growing one or more vegetable crops on a large scale for fresh shipment to distant markets. Most important truck crops are Potato, tomatoes, lettuce, melons, beets, broccoli, celery, radishes, onions, cabbage, and strawberries.

(5) Classification Based on Cultural Method/Water

(i) **Rain fed:** Crops grow only on rain water, e.g Jowar, Bajra, Mung etc.

(ii) **Irrigated crops:** Crops grown with the help of irrigation water, *e.g*,. Chili, sugarcane, Banana, papaya etc.

(6) Classification Based on Root System

(i) **Tap root system:** The main root goes deep into the soil. *e.g.*, Tur, Grape, Cotton etc.

(ii) **Adventitious/Fiber rooted:** The crops whose roots are fibrous shallow & spreading into the soil. *e.g.*, Cereal crops, wheat, rice etc.

(7) Classification Based on No. of Cotyledon

(i) **Monocots or monocotyledons:** Having one cotyledon in the seed. *e.g.*, all cereals & Millets.

(ii) **Dicots or dicotyledonous:** Crops having two cotyledons in the seed. *e.g.*, all legumes & pulses.

CHAPTER 9

National and International Agricultural Setup

Majority of people which is almost half of the world's population live in rural area. Around 2.5 billion peoples derive their livelihoods from agriculture. Many poor developing countries, primary activities such as agriculture still constitute the backbone of the economy. Ample scope and opportunity for agriculture sector is that almost 12.5 percent of the global population is under nourished. The agriculture sector does not provide a large part of global gross domestic product which continues to be dominated in all regions by services with almost 70 percent share followed by about 20 per cent in industry and agriculture contributes about 5 per cent. However, agriculture plays an important role, especially because approximately one third of the world's population still survives its livelihood from agriculture. In developing countries its share is certainly higher in the range of 14 to 50 %.

Indian agriculture has performed remarkably well in terms of commodity growth. Indian agriculture contributes to 8% global agricultural gross domestic product to support 18% of world population on only 9% of world's arable land and 2.3% of geographical area. The growth and development of agriculture has to be viewed minutely of research and development since the primary stake holders of the sector viz. farmers are the determinants and decide the very survival of any such system. The foundations for the organized contributions for research and technology to agriculture were probably laid in with the starting of agriculture departments in the states followed by higher education colleges, famine commission, the Imperial Agricultural Research Institute (1905), Royal Commission on Agriculture (1926) and so on till the latest World Bank assisted National Agricultural Innovation Project. Several institution promoted by government, corporate and NGOs have come up in the past century or so that set the mandate to conduct research and development leading to technologies for sustainable agriculture. These institutions although might be owned by different agencies but have the common broad agenda to generate and transfer technologies for the better of Indian agriculture. The Indian Council for Agriculture Research has taken upon itself the onus of coordinating and driving these institutions forward. Therefore, the future of agriculture and the agriculturists has strong bond with National Agricultural Research System (NARS).

International Agricultural Research System

The global agricultural research system is driven by the Consultative Group on International Agricultural Research (CGIAR). CGIAR is a global partnership that unites organizations engaged in research for a food secure future. CGIAR research is focused on reducing rural poverty, increasing food security, improving human health and nutrition, and ensuring sustainable management of natural resources. CGIAR was initiated in the year of 1971; it had only four Centers and eight members, representing national research systems. At present agricultural research is carried out by 15 of CGIAR Centers that are members of the CGIAR Consortium along with 64 country members and it works in close collaboration with hundreds of partners, including national and regional research institutes, civil society organizations, academia, development organizations and the private sector. The CGIAR Fund is a multi-donor trust fund that finances CGIAR research guided by the Strategy and Results Framework. The CGIAR fund is administered by the World Bank, as Trustee, and governed by the Fund Council, a representative body of Fund donors and other stake holders. The CGIAR Consortium is an international organization that, together with the CGIAR Fund, advances international agricultural research for a food secure future by integrating and coordinating the efforts of those who fund research and those who do the research. The CGIAR Consortium is made up of the Consortium Board; The Consortium Office; and the Research Centres which are members of the CGIAR Consortium. The following are the 15 CGIAR Institutes or Centres are given below:

1. Africa Rice Centre
2. Bioversity International
3. Centre for International Forestry Research (CIFOR)
4. International Centre for Agricultural Research in the Dry Areas (ICARDA)
5. International Centre for Tropical Agriculture (CIAT)
6. International Crops Research Institute for the Semi-Arid Tropics (ICRISAT)
7. International Food Policy Research Institute (IFPRI)
8. International Institute of Tropical Agriculture (IITA)
9. International Livestock Research Institute (ILRI)
10. International Maize and Wheat Improvement Centre (CIMMYT)
11. International Potato Centre (CIP)
12. International Rice Research Institute (IRRI)
13. International Water Management Institute (IWMI)
14. World Agroforestry Centre (ICRAF)
15. World Fish

National Agricultural Research System

The prominence of agricultural research was realized globally due to frequent weather abnormalities like famines and epidemics. It was probably based on this model that the famine commission recommended setting up of state departments of agriculture in India and much later i.e. almost after once century India adopted the land grant scheme to start Agriculture University. Institutions with Union and state support were started across regions and covering various systems and commodities. Thus, the agricultural research in India has built up a fairly advanced system with good network. The Indian NARS is one of the largest systems in the world. The effective functioning of this system, in close association with education and extension systems has greatly contributed to the rapid growth of agriculture after independence. The ICAR and the national agricultural research system have played a pioneering role in ushering green Revolution and subsequent developments in agriculture in India through its research and technology development that has enabled the country to increase the production of food grains by 4 times, horticultural crops by 6 times, fish by 9 times (marine 5 times and inland 17 times), milk 6 times and egg 27 times since 1950-51, thus making a visible impact on the national food and nutritional security. It has played a major role in promoting excellence in higher education in agriculture. It is engaged in cutting edge areas of science and technology in their fields.

India has one of the largest agricultural research systems in the world with the largest number of scientific personnel of any developing country engaged in research and education relating to agriculture and allied areas. The research system includes approximately 30,000 scientists and more than 100,000 supporting staff actively engaged in research related to agriculture. Although the total number of scientists engaged in agricultural research in India looks very impressive, it compares less favorably with many developed nations in the world. The research system has evolved over years of innovation and experimentation, and it has withstood the test of time remarkably well. The present agricultural research comprises essentially two main streams, viz. the ICAR at the national level and the Agricultural Universities at the state level. Besides, several other agencies such as the Conventional / General Universities, Scientific Organizations and various Ministries or departments at the Centre, and also Private or Voluntary Organizations participate directly or indirectly in research activities related to agriculture. ICAR is a major scientific organization in the country, having on current responsibility for both research and education. As an apex body at the national level, ICAR is mainly responsible for the promotion and coordination of agricultural research in the various branches of agriculture and allied sciences in the country. In addition to its promoting and coordinating roles, ICAR is also directly involved in undertaking research at the national level, basic as well as applied, on diverse problems facing production of crops animals, fisheries, etc., with the objective of evolving new production technologies suited to different agro-climatic conditions. Just as the University Grants Commission (UGC) plays a major role for the general education in the country, ICAR plays a similar role in the area of agricultural education. The Charter of the ICAR also includes extension education, which is carried out through a network of projects and other mechanisms.

Organizational Structure

The ICAR is an autonomous organization under the Department of Agricultural Research and Education (DARE), Ministry of Agriculture, Govt. of India. Formerly known as

Imperial Council of Agricultural Research, it was established on 16th July, 1929 as a registered society under the Societies Registration Act, 1860, in pursuance of the report of the Royal Commission on Agriculture. The Council is the apex body for coordinating, guiding and managing research and education in agriculture including horticulture, fisheries and animal sciences in the entire country. With 102 ICAR institutes (including four deemed universities) and 65 Agricultural Universities spread across the country this is one of the largest national agricultural system in the world.

Research Infrastructure of ICAR

ICAR has established many Central Research Institutions over the years to meet the agricultural research needs of the country. They are given below-

- Deemed Universities-4
- ICAR Institutions-65
- National Bureaux-6
- Project Directorates-13
- National Research Centres-14
- Substations of ICAR Institutes-138
- AICRPs (All India Coordinated Research Projects)-61
- Other Projects-10
- Network Projects-19
- Zonal Project Directorates-8
- State Agricultural Universities (SAUs)-64
- Central Agricultural University -3
- Central Universities having faculty of agriculture-4
- Krishi Vigyan Kendras (KVKs) - 706
    - Zone I – 69 KVKs
    - Zone II – 63 KVK
    - Zone III – 83 KVKs
    - Zone IV – 63 KVKs
    - Zone V – 59 KVKs
    - Zone VI – 46 KVKs
    - Zone VII – 43 KVKs
    - Zone VIII – 79 KVKs
    - Zone IX- 79 KVKs
    - Zone X-74 KVKs
    - Zone XI- 48 KVKS

Headquarters

Indian Council of Agricultural Research, New Delhi

Externally Funded Projects in India

1. National Agricultural Research Project (NARP)

Agricultural Universities, which have a state-wide mandate for agriculture, did not have a strong base for research at the regional level. Most of the funds provided were utilized for developing the University main campuses, thus neglecting the regional research needs. To overcome this, the ICAR launched in 1979, with World Bank assistance, a novel scheme known as National Agricultural Research Project (NARP) to strengthen the regional research capabilities of these Universities for conducting need – based, location specific and production oriented research in identified agro-climatic zones. Under this project, each State is divided into a contiguous set of agro-climatic zones on the basis of climate, soils, crops and ecology. There are 131 such zones in the country with 125 zones in 17 states and Union territories and 6 zones in the North Eastern Region. In each zone, a major regional research station with a multidisciplinary team of scientists is established or strengthened. The strengthening is accomplished by providing funds for incremental scientific and supporting staff, laboratory, equipment, transport, operating cost and other physical facilities.

The Agricultural Universities, on its part, has to provide funds for land, cultivation and station maintenance. The project also provides for strengthening the office of the Director of Research in these Universities for coordinating the research activities undertaken by these stations more effectively. An inventory of resources and systems of cultivation are prepared for each zone to serve as a basic document known as 'Status Report' for identifying regional research needs. Farming systems research with emphasis on multidisciplinary approach to problem solving and establishment of a closer linkage between research and extension at the grassroots level with active participation of farmers are the essential components of this project. The project is guided and serviced by the NARP Directorate located in the ICAR headquarters. The progress of the project is regularly monitored by the Agricultural University and the ICAR and also is reviewed periodically by the special Mission of the World Bank. The ICAR's assistance is available for a period of five years after which the responsibility for continuing the project rests with the University. The project had the IDA support of $ 27 million in the first phase, which was about 50% of the project cost. The first phase laid major emphasis on cereals, pulses and oilseeds under rain fed conditions. In this phase, nearly 109 sub- projects were approved and it was closed in September 1985 after having used $19.5 million of the $27 million credit. The second phase on NARP, under implementation since February 1986, is intended to intensify and continue the process of decentralizing agricultural research. In addition to financing the sub-projects not completed under the first phase, this phase provides support for special research projects in irrigated farming, animal drawn equipment, horticulture including post-harvest technology, commercial crops, agroforestry and animal nutrition. On the recommendation of the mid term review team, the second phase of the project was extended up to 1994. Strengthening of regional research on a scientific agro-climatic basis has been the most significant positive development that has taken place through the implementation of NARP in the organization of agricultural research in the country. In spite of some minor problems, the project has achieved its main objective of helping to strengthen the regional research capabilities of the Agricultural Universities and decentralize agricultural research. The project has created an awareness of developing mission-oriented, problem-specific, relevant research with multidisciplinary thrust and thus, has made considerable impact in many areas where enough attention was not paid earlier.

2. National Agricultural Technology Project (NATP)

With a view to raising the efficiency of resource use for technology generation and assessment as well as transfer, i.e. involving both agricultural research and extension, ICAR has initiated the major National Agricultural Technology Project (NATP) in 1998 with the financial support provided by the World Bank to the tune of US$ 243 million. After successful completion of five years, the project came to an end in 2004. The project essentially focused on the following components: Organization & Management (O&M) Reforms Agro-ecosystem Research; Innovations in Technology Dissemination; Competitive Grants Scheme; of these, Agro-ecosystem research constituted the major activity. This component concentrated on the following areas: Location- specific Production Systems Research (PRS) in the programme mode; Cross-cutting Mission Mode Research to support location-specific PSR; Teams or Centers of excellence to support strategic research related to PSR. The NATP gave fillip and impetus on productions systems approach the themes.

3. National Agriculture Innovation Project (NAIP)

Agriculture is and will continue to be the main driver of country's economic growth with social justice. Our agriculture did extremely well and it was on the ascendancy till the mid-nineties but after that the growth slowed down. Since 1996-97, the growth rate of agricultural GDP has been, on an average, 1.75 % per year in contrast with the rate of 4% that is required. On the other hand the farmer has been facing rising input costs, declining returns from inputs, uncertain market, increasing role of market in agriculture and blurring of distinction between the domestic market and the international market. To assist the farmers in these changing contexts new strategies and innovative solutions are urgently required which in turn will require technological support. Hence, the agricultural research system which generates technologies has to conduct the business of agricultural research in an innovative way. The World Bank aided National Agricultural Innovation Project (NAIP) has been convinced to pilot this innovation in conducting agricultural research. The basic principles of National Agricultural Innovation Project are stated following:

(a) To give the agricultural research and technology development system an explicit development and business perspective through innovative models. In other words, the agricultural research system should be able to support agriculture as a business venture and also as a means of security of livelihood of the rural Indian while maintaining excellence in science.

(b) To make the national Agricultural Research System a 'pluralistic' system where every Organization having stake in agricultural research: public private or civil society, has to play a role.

(c) Working in well defined partnership groups with clear common goals and understanding on sharing responsibilities and benefits.

(d) Funding through competition so that a wide choice of excellent innovative ideas come in from the stakeholders themselves.

(e) Work with focus, plan and time frames.

Develop well tested models for application of agricultural research and technology for profitability of farming, income generation and poverty alleviation.

Components of National Agricultural Innovation Project (NAIP)

The NAIP will function through four components:

(a) The ICAR as the Catalysing Agent for the Management of Change in the Indian NARS

(b) Research on Production to Consumption Systems (PCS)

(c) Research on Sustainable Rural Livelihood Security (SRLS) and

(d) Basic and Strategic Research in the Frontier Areas of Agricultural Sciences (BSR)

The Agricultural Universities System

As agriculture is a state concern, the responsibilities for research, education and extension rest with the State Governments. Prior to 1960, agricultural research in the States, essentially on local problems, was carried out by the State Departments of Agriculture. During the past 40 years, research and education have been transferred to the Agricultural Universities and the State Departments of Agriculture organize extension services. The Universities are supported by their respective State Governments. ICAR provides financial support and assists their research and education programmes. The University Education Commission (1949) recommended the setting up of 'Rural Universities'. This was endorsed by the two Joint Indo-American Teams in 1955 and in 1959, as well as the Ford Foundation Study Team in 1959. In 1960, the Agricultural Universities Committee constituted under Chairmanship of Dr. Ralph W. Cummings prepared certain guidelines for the establishment of Agricultural Universities in different States, and the ICAR gave necessary support. The first Agricultural University was established at Pant Nagar in Uttar Pradesh in 1960, patterned on the Land-Grant System of the United States. The second Education Commission (1964-66) recommended at least one Agricultural University in each State and ICAR prepared a model Act in 1966. All the States have now at least one Agricultural University each. Though the Model Act specifies that only one University shall be established in each State, which was later endorsed by the National Commission on Agriculture, many States have established multiple Universities to meet the regional needs. There are at present 62 State Agricultural Universities, 5 Deemed Universities (4 in ICAR, one SHIATS formerly AAI, Naini), two Central Agricultural University in North Eastern Region and in Jhansi and 4 Central Agricultural University with agriculture faculties. These include Veterinary and Animal Science Universities, in some of the States. Some Agricultural Universities, as in Maharashtra State, have affiliated private agriculture colleges.

Research Infrastructure in State Agriculture University

Basically, the research infrastructure consists of an Experiment Station at the main campus and a number of Regional Research Stations and Substations located in different parts of the state. There are over 500 such Research Stations belonging to the agriculture Universities, working on location specific problems. Generally, the research programmes are headed by the Directors of Research, who are assisted by the Associated Directors of Research located at the Regional Research Stations with in the state. Some Agricultural Universities have established Advanced Centres by combining related subjects in areas such as plant protection, genetics and Plant breeding, agricultural engineering, agricultural economics, water technology, etc, In order to undertake need based and location specific research, a network of Zonal Agricultural Research and Extension Centres have been established since 1979 with assistance from the World Bank under NARP.

General Universities and Agriculture Linkup

Many General Universities with well developed faculties in agriculture, or strong departments engaged in areas such as genetics, plant physiology, mycology, entomology, biochemistry, economics, chemistry, marine biology, home science, etc. have made distinctive contributions to agriculture in the country. Besides, the Central Universities like Banaras Hindu University, Visva Bharati University, etc. have institutes of Agricultural Sciences, which are engaged in research in agriculture and allied areas, some of which are supported by the ICAR.

Other Scientific Organizations and its Agricultural Activity

Many other scientific organizations either directly undertake research, or sponsor or support programmes related to agriculture. The Council of Scientific and Industrial Research (CSIR), through its network of national Laboratories provides research support in areas like processing of agricultural products, recycling of agricultural wastes, development of various agro-chemicals, etc. The Indian Council of Medical Research's (ICMR) research on the nutritional qualities of the various agricultural produce including toxicity and occupational health of agricultural workers have greatly helped the ICAR in planning its research programmes. Some of the areas in which Bhabha Atomic Research Centre (BARC) is actively engaged are the development of the newer varieties of crops and preservation of agricultural produce. The Indian Space Research Organisation (ISRO) is helping the research system to assess Indian's soil and water resources. Technological institutions like IIT, Kharagpur, are active in the fields of agricultural engineering, soil and water management, agronomy. The Department of Science and Technology (DST) and Department of Biotechnology (DBT) promote research on genetic engineering, post-harvest technology, and areas of basic sciences supportive to agriculture. The Department of Non-Conventional Energy Sources works on the utilization of solar and wind energies and biogas for agricultural purposes. The Department of Meteorology is actively engaged in research on weather forecasting. The Department of Ocean Development is involved in assessing the fishery resources in the country and promotes research in the area of fisheries. Similarly the ministry of textiles through Central Silk Board promotes research on sericulture and silk technology with multi-location presence. Institutions like the National Dairy Development Board (NDDB) under the Agriculture Ministry; various Commodity Boards like Coffee, Rubber, Tea and Cardamom Boards under the Commerce Ministry; and the Forest Research Institute and Wasteland Development Board under the Ministry of Forestry and Environment help in strengthening the agricultural research system in the country.

Private Agriculture Sector

Involvement of private sector in agriculture research is of considerable importance in the overall development of agriculture in the country. In mid 1960s, several private companies started programmes mainly to develop hybrid maize, sorghum and bajra. Research on vegetables started in late 1960s. Private sector research is confined mainly to breeding crop hybrids, certain plantation crops, agro-chemicals, poultry and agricultural machinery. Private sector research in seed industry has gone very rapidly. Several private companies are now engaged in the production of hybrid seeds of a variety of crops like cotton, sorghum, bajra, maize, vegetables, red gram, rice, etc., and a small seed export industry has also emerged. There are a number of private companies undertaking plant breeding research, and several others are involved in plant protection research. Besides their own research stations, these companies conduct experiments on farmers' fields. They test the bio-efficacy of insecticides and herbicides that are new to India as well as synthesize new compounds.

Private research in poultry sector is of paramount importance. Although Government introduced exotic birds in commercial poultry industry, they were popularized by the private sector and many companies are now actively engaged in it. Many other large industrial concerns are engaged in research on shrimps and shrimp feed. Some private companies undertake major research and development programmes for the improvement of tractors and irrigation pumps. Research on tractors seems to be primarily aimed at improving quality, fuel efficiency and engine durability. Some companies are now moving into agricultural implements. Research in pump industry is aimed at increasing the efficiency of pumps through improved design and better materials. Some companies are even experimenting with non-conventional sources of power. Historically, private companies in the processing and plantation sector have been a very important source of new agricultural technology. Some of the prominent ones include Indian Sugar Mills Association, Southern Planters Association, Textile Mills Association, Silk Industry, etc. Some large farms involved in research on animal nutrition, plant growth regulators, biotechnology like tissue culture in cardamom, sugarcane, coconut and tea, bio-fertilisers, etc. Research in the area of tree farming including in vitro culture and tree breeding is also receiving attention of the private firms. Some of the well-established institutions such as Allahabad Agricultural Institute, Bharatiya Agro-Industries Foundation, Wool Research Association, and United Planters Association of South India undertake short-term, mission-oriented research projects supported by the ICAR for multi-location testing of varieties and agro-techniques. In order to promote scientific research and the participation of industry in it, the Government through the Income Tax Act of 1961 has offered certain tax concessions relating to the expenditure on scientific research. The involvement of private agencies in agricultural research is gaining momentum with greater sophistication in technological development and the prospects of high returns on investment in agriculture.

Linkages and the Subsystems

Significant working relationships and complementarily in research efforts amongst the components of the research system is necessary in order to optimize resources and check avoidable duplication. The ICAR, as the coordinating agency at the national level, has established close-working relationships with the Agricultural Universities and other agencies involved directly or indirectly in agricultural research through formal arrangements and informal exchanges. At the policy level, the Vice-Chancellors of Agricultural Universities are represented in the Governing Body, and in the Norms and Accreditation Committee of the ICAR. The senior level research managers of the ICAR in turn, are represented in the Management Boards of these Universities. The Regional Committees of the ICAR provide an important forum for the scientists from these two agencies to come together and look at the regional research needs. Through Interdisciplinary Scientific Panels of the ICAR, the experts from the Agricultural Universities play a critical role in selecting research programmes at the national level as well as at the regional level. More importantly, various research schemes of the ICAR like the AICRPs, NATP, and ad hoc research schemes provide opportunities for the two subsystems to work jointly on problems of national as well as regional relevance.

As far as the General Universities are concerned, they participate in research activities under different types of research schemes and projects financed by different agencies. Through the AICRPs and ad hoc research schemes, these Universities have established linkages with the ICAR and agricultural Universities subsystems. Joint programmes in specific areas like plant physiology, biological nitrogen fixation, biotechnology, etc. have been taken up by the ICAR with scientists working in these Universities. ICAR has also established close linkages with various scientific organizations like CSIR, ICMR, ISRO, BARC, etc. through Joint Panels. Problems of mutual interest have brought the ICAR closer to various Departments and Ministries at the Centre to find solutions through collaborative research efforts.

International Agricultural Co-operation

International co-operation has played a important role in developing and strengthening the academic and research system in India. Many countries like USA, UK, USSR, Canada, Australia, Japan, several European countries; Charitable Institutions, etc.; Rockefeller and Ford Foundation; various International Agencies like FAO, UNDP, UNESCO, World Bank, etc.; and the International Agricultural Research Centres under the Consultative Group on International Agricultural Research (CGIAR) System have contributed extensively to the cause of agricultural research in India. Spectacular achievements in increasing the food production have raised the country's development considerably. The reciprocity and mutuality of interests with the less developed and as well as the technologically advanced countries are the essence of international co-operation. The Government has authorized ICAR, assisted by the DARE, to enter into bilateral cooperative agreements with several countries and agencies. The mode of collaboration normally follows the pattern of: (a) exchange of germplasm of plant and animal origin; (b) exchange of scientific and technical information; (c) visits of scientists and experts; (d) training of scientists; and (e) infrastructure development.

CHAPTER 10

Current Scenario of Indian Agriculture

Agriculture is the mainstay of Indian Economy and about 65% of Indian population depends directly on agriculture. It has to support almost 17 per cent of world population from 2.3 per cent of world geographical area and 4.2 per cent of world's water resources. During the year 2018, a total of 372 high yielding varieties/hybrids comprising 200 of cereals, 49 oilseed, 47 pulses, 47 commercial crops and 29 of forage crops were released. In the horticultural crops, 120 varieties have been developed in vegetables, fruits, spices and tuber crops, A total of 24 new 24 breeds of livestock and poultry were registered. The total number of registered indigenous breeds in the country to 184 including 43 of cattle, 16 of buffalo, 34 of goat, 43 of sheep, 7 of horses and ponies, 9 of camel, 8 of pig, 2 of donkey, 1 of yak, 19 of chicken, 1 of duck and 1 of geese. The nation is striving to find ways and means to keep its burgeoning population adequately fed. The available natural resources and existing infrastructure is the only way to make the ends meet. Management of the already built infrastructure in harmony with natural systems is the clarion call of the day. Knowledge of the extent of existing infrastructure and natural resources is one of the most basic pre-requisites to utilize them effectively and in a sustainable manner. Efforts were made to promote zero budget farming and other indigenous agricultural technologies (ITKs) to control pest population using pheromone trap in vegetables and fruits.

During the year of 2017-18, 1.77 lakh quintal seeds of improved varieties and hybrids of cereals, oilseeds, pulses, commercial crops, vegetables, flowers, fruits, spices, fodder, forest species, medicinal plants and fibres crops, were produced and provided to 19.98 lakh farmers. Bio-agents (70.96 q), bio-pesticides (381.70 q), bio-fertilizers (12,307.17 q), vermin-compost, mineral mixture etc. were produced and supplied to 3,890 farmers.

The report on the state of Indian agriculture placed by Government to Rajyasabha, calls for wide ranging reforms in agriculture sector to enable it to meet the growing demands and meet the challenges posed by various human and environmental factors. endeavours to develop technologies for enhancing productivity and reducing the cost of cultivation. In recent years, agriculture sector in India has witnessed spectacular advances in the production and productivity of food grains, oilseeds, commercial crops, fruits, vegetables, food grains, poultry and dairy. Adoption of high yielding varieties by farmers coupled with the use of higher doses of fertilizer and assured irrigation through tube wells accelerated the pace of progress in agriculture. As a result of adoption of improved inputs and management practices, the total food grain production increased from a mere 50.8

million tonnes in 1950–51, to 212 million tonnes in 2006–07 and productivity increased from 522 kg/ha to more than 1707 kg/ha. The productivity of wheat, rice and oilseeds increased to a greater extent than other crops.

The increase in production of food grain was possible as a result of adoption of quality seeds, higher dose of fertilizer and plant protection chemicals, coupled with assured irrigation. India has emerged as the second largest producer of fruits and vegetables in the world in addition to being the largest overseas exporter of cashews and spices. Further, India is the highest producer of milk in the world. India has Monsoon climate in which a year has been divided into two distinct seasons of summer and winter. Rainfall occurs mainly in summer. India has a strong weather forecasting system developed and maintained by Indian Meteorological Department (IMD). Apart from weather forecasting and severe weather warning, it also gives agro-meteorological services to farmers in India. India has diverse agro-climatic zones from north to south and from east to west. It has been divided into fifteen different agro-climatic zones, which signifies its diversified agricultural production from tropical to temperate crops.

Rice, Wheat, Sugarcane, Oilseeds, Pulses, Cotton, Jowar, Bajra, Ragi, Tea, Coffee, Coconut, Cashew, Rubber, Spices, Cauliflower, Onion, Cabbage, Mango, Banana, Sapota, Acid lime. Indian Agriculture is characterized by small and marginal operational holdings. About 85 % of total cultivated land has been fragmented into less than 10-hectare land. About 60 % of farmland is less than 4 hectare in size. India ranks first in the world in milk production, which was around 100 million tonnes in 2016–17. Strong networks of Milk Cooperatives have been instrumental in this phenomenal performance of dairy sector in India. Presently, 1.13 lakh village level cooperative societies spread over 265 districts in the country form part of the national Milk Grid. This Grid links milk producers throughout India and consumers in 700 towns and cities. Livestock sector contributes about 27 % of the G.D.P. from agriculture and allied activities. This sector has excellent forward and backward linkages, which promote many industries and increase the incomes of vulnerable groups of the society, such as agricultural labourers and small and marginal farmers. Livestock plays an important role in the socioeconomic life of India. It is a rich source of high quality foods, such as milk, meat and eggs and a source for income and employment to millions of rural farmers, particularly women. With a large human population and more than 250 million economically strong potential consumers, the domestic demand for these food products are increasing rapidly, the demand often exceeding the supply. India possesses the second largest livestock population in the world. Production and export of poultry products have shown considerable growth in the recent decades. Export of such products to countries including Bangladesh, Srilanka, Middle East, Japan, Denmark, USA, and Angola augers well for this industry. Fishing, aquaculture and a host of allied activities are a source of livelihood to over 14 million people and a major source of foreign exchange earner. About 8,118 k.m. of coastline gives geographical basis for the development of marine fishery sector and cultural factor boosts the inland fishery sector in India.

Agriculture is a critical sector of the Indian economy. An average Indian spends almost half of his/her total expenditure on food, while roughly half of India's work force is still engaged in agriculture for its livelihood. Being both a source of livelihood and food security for a vast majority of low income, poor and vulnerable sections of society, its performance assumes greater significance in view of the proposed National Food Security Bill and the ongoing Mahatma Gandhi National Rural Employment Guarantee Act (MGNREGA) scheme. The experience from RICS countries indicates that a one percentage growth in agriculture is at least two to three times more effective in reducing poverty than the same growth emanating from non-agriculture sectors. Given that India is still home to the largest number of poor and malnourished people in the world, a higher priority to agriculture will achieve the goals of reducing poverty and malnutrition as well as of inclusive growth. Since agriculture forms the resource base for a number of agro- based industries and agro-services, it would be more meaningful to view agriculture not as farming alone but as a holistic value chain, which includes farming, wholesaling, warehousing, processing, and retailing.

Further, it may be noted that in the last two Five Year Plans, it is clearly mentioned that for the economy to grow at 9 per cent, it is important that agriculture should grow least by 4 per cent per annum. Achieving an 8-9 percent rate of growth in overall GDP may not deliver much in terms of poverty reduction unless agricultural growth accelerates. At the same time 'growth with inclusiveness' can be achieved only when agriculture growth accelerates and is also widely shared amongst people and regions of the country. All these factors point to just one thing: that agriculture has to be kept at the centre of any reform agenda or planning process, in order to make a significant dent on poverty and malnutrition, and to ensure long-term food security for the people.

Major Problems of Indian Agriculture

The major problems faced by the Indian agriculture are those of population pressure, small holdings, depleted soils, lack of modern technology and poor facilities for storage.

(a) Population Pressure

India has a huge population of over one billion and it is increasing at a very fast rate. According to 2001 census figures the overall density of population is 324 persons per sq. km. This is likely to increase further in future. This has created great demand for land. Every bit of land has been brought under the plough. Even the hill slopes have been cut into terraces for cultivation.

(b) Small and Fragmented Land Holdings

The pressure of increasing population and the practice of dividing land equally among the heirs has caused excessive sub divisions of farm holdings. Consequently, the holdings are small and fragmented. The small size of holdings makes farming activity uneconomical and leads to social tension, violence and discontentment.

(c) Inadequate Irrigation Facilities

By and large the irrigation facilities available in India are far from adequate. So for half of the total area under food crops has been brought under irrigation and the remaining half is left to the mercy of monsoon rains which are erratic in time and space.

(d) Depleted Soils

Indian soils have been used for growing crops for thousands of years which have resulted in the depletion of soil fertility. With deforestation the sources of maintaining natural fertility of soil has been drying out. Lack of material resources and ignorance of scientific knowledge have further depleted the soils of the natural fertility. Earlier only animal waste was enough to maintain soil fertility.

(e) Storage of Food Grains

Storage of food grains is a big problem. Nearly 10 per cent of our harvest goes waste every year in the absence of proper storage facilities. This colossal wastage can be avoided by developing scientific ware-housing facilities. The government has taken several steps to provide storage facilities.

(f) Farm Implements

Although some mechanization of farming has taken place in some parts of the country, most of the farmers are poor and do not have enough resources to purchase modern farm implements and tools. This hampers the development of agriculture.

Structure of Indian Agriculture

The agriculture sector in India has undergone significant structural changes in the form of decrease in share of GDP and State of Indian Agriculture indicating a shift from the traditional agrarian economy towards a service dominated one. This decrease in agriculture's contribution to GDP has not been accompanied by a matching reduction in the share of agriculture in employment. About 52% of the total workforce is still employed by the farm sector which makes more than half of the Indian population dependant on agriculture for sustenance. The average size of operational holdings in India has diminished progressively from 2.28 ha in 1970-71 to 1.55 ha in 1990-91 to 1.23 ha in 2005-06. As per Agriculture Census 2005-06, the proportion of marginal holdings (area less than 1 ha) has increased from 61.6 percent in 1995-96 to 64.8 percent in 2005- 06. This is followed by about 18 percent small holdings (1-2 ha.), about 16 percent medium holdings (more than 2 to less than 10 ha.) and less than 1 percent large holdings (10 ha. and above). With the declining share of agriculture to GDP, the continuing high pressure of population on agriculture and the increasing fragmentation of land holdings leading to decreasing availability of cultivated land area per household, the agriculture sector alone would hardly be in a position to create additional employment opportunities. This fragmentation of land has serious consequences in terms of owning and using modern technology and farm equipments, and also in provision of inputs including agricultural credit and extension. The consequences are high transaction cost for farmers, banks and service providers, inability to own or finance farm equipment because of fluctuating incomes of tiny farmers and viability considerations, smaller volumes of produce reducing ability to negotiate at market place etc. Higher cost of cultivation and lower incomes are a serious disincentive for farmers to continue farming. There is a need to find ways that enable these small ownership holdings aggregate to minimum viable size of operating farms without the individual landowners losing their ownership rights. Co-operative farming, collective farming, producers organizations, Joint Liability Groups (JLGs), leasing out land, or contract farming are some possible ways of aggregation. While the first three solutions are fraught with capital and managerial issues, and JLGs offer perhaps a temporary and an intermediate solution, a long-term solution can be seen only in leasing out land to individuals and in contract farming where the contract is evenly balanced. Unfortunately, the present lands laws in most states are based on land reforms carried out in the sixties and seventies, and were based on the maxim 'land to the tiller'. Leasing out agricultural land is not permitted in most of the major agricultural states such as Bihar, Gujarat, Jharkhand, Karnataka, Kerala, Madhya Pradesh, Odisha, Uttar Pradesh, Telangana area of Andhra Pradesh, Himachal Pradesh, Jammu and Kashmir, etc. except by Defence personnel, minors, disabled persons and the Prakash Bakshi, "Whither Indian Agriculture", This calls for creation of additional employment opportunities in the non-farm and manufacturing sector, especially in agro based rural industries which have area specific comparative advantage in terms of resources endowment and development possibilities. This would require suitable skill development of the people so as to gainfully employ them in non-farm activities. This alone would be able to make agriculture viable in a sustainable manner. In addition, by creating more employment and absorbing some of the surplus labour in agriculture, this will contribute to achieving our objective of inclusive growth.

Fragmentation of operational holdings has widened the base of the agrarian pyramid in most states. Empirical studies have, however, demonstrated that agricultural productivity is size neutral. Factors that determine productivity favourably include among others an easy and reliable access to modern inputs, access to suitable technology tailored for specific needs, the presence of support infrastructure and innovative marketing systems to aggregate and market the output from such small holdings efficiently and effectively. In agricultural technology, the use of high yielding varieties as in the case of Bt cotton and maize, economy in input use, the availability of quality seeds and farming techniques such as system of rice intensification enabled finally by marketing links all have high potential to improve yield. The growth performance of the agriculture sector has been fluctuating across the plan periods. It witnessed growth rate of 4.8 per cent during the Eighth plan period (1992–97). It has had some success in that food grain production touched a new peak of 250.42 million tonnes in 2011-12. Agricultural GDP growth has accelerated to an average 3.9 percent growth during 2005- 06 to 2010-11. The increasing divergence between the growth trends of the total economy and that of agriculture and allied sectors suggests an under performance by agriculture. It is also significant that unlike the overall economic growth pattern, agricultural performance in India has been quite volatile. This is almost six times more than the CV observed in the overall GDP growth of the country indicating that high and perhaps increasing volatility is a real challenge in agriculture, which is likely to increase in the years to come in the wake of climate change. The Indian agriculture growth pattern has been highly varied at the state level. Since agriculture is a state subject, the overall performance of the agriculture sector in India largely depends on what occurs at the state level. There is a wide variation in the performance of different states.

Growth in the production of agricultural crops depends upon acreage and yield. Given the limitations in the expansion of acreage, the main source of long-term output growth is improvement in yields. Rice production achieved during 2016-17 is 210 million tones and wheat production 98 million tones. Coarse cereal production during 2016-17 is 44 million tones and food grains production 275 million tones. Pulses, oilseeds, sugarcane, cotton and jute and mesta production achieved during 2016-17 was 22.95, 32.10, 306.72, 33.06 and 10.60 million tones respectively. This suggests the need for renewed research to boost production and productivity. In pulses, gram recorded a growth of 6.39 percent in production during the same period driven by expansion in the area under cultivation. Soya bean has recorded a high rate of growth in production in both the periods, driven primarily by expansion in area under cultivation. In fact oilseeds as a group have shown some significant changes in the two decades: the production growth rate has more than doubled in the decade of 2000s over the previous decade, driven both by productivity gains as well as by area gains.

Currently food grains constitute about one fifth of the total value of output from the agriculture & allied sector which is less than the contribution from the livestock sector and almost equal to that of the horticulture Sector. The shares of fruits & vegetables and livestock have shown an increasing trend in recent years implying that they have been growing at a much faster rate than the traditional crops sector. Given the rising share of high value commodities in the total value of agricultural output and their growth potential, this segment is likely to drive agricultural growth in the years to come. Being highly perishable in nature, this segment requires faster and better linkages between farms and firms in terms of logistics, processing and organized retailing. This would entail institutional changes that can incentives entrepreneurs to invest in building efficient and faster value chains that reduce wastages, and increase the incomes of the farmers at the bottom of the chain.

Status of Indian Agriculture

Indian agriculture is the largest private sector enterprise in the country of over 110 million farmers. It engages two-third of total workforce, contributes to 26 per cent of the GDP and nearly one-sixth of the total export earnings. India inherited a stagnant agriculture at the time of independence growing at an annual rate of growth of 0.3 per cent in the first half of the last century. Due to the planned Government policies, technological support provided by agricultural scientists and hard work put in by the farmers, Indian agriculture has achieved an annual growth rate of 2.7 per cent in the post independence era. This revolution not only became the national pride as crusader against hunger and poverty but also a model for the developing world. Nevertheless, it is time to work to produce more diversified and quality food at affordable price to meet the requirements of the ever increasing population and also improve the living standards of people. This must draw our serious attention now. It may be worth sharing that during the last three decades, the production increase was ushered by use of high yielding varieties of seed, assured supply of water and intensive use of chemical fertilizers and pesticides. Such a level of chemical fertilizer intensity may be undesirable for both environment and productivity reasons. Similar question may bother our minds for other inputs.

The factors responsible for low productivity could largely be, research breakthrough in restricted few crops, growing disparity between potential yields due to improper technology transfer mechanism, low productivity from rainfed crops, low input use and also low input use efficiency, lack of food processing and value addition to horticultural and vegetable crops, high post harvest losses especially in horticultural produce, very low agricultural produce export; poor market infrastructure, untrained human resource in agriculture sector especially as entrepreneurs and skill workers. Special efforts need to be made to raise the productivity and production of crops, livestock, fisheries, horticultural crops to ensure food and nutritional security for the people and generating surplus for export keeping in view the fact that India has a great competitive advantage in agricultural export.

Sustaining the Development

The rapid change in Indian Agriculture particularly since the era of green revolution has been received with mixed response. The development perspective has been definitely encouraging and viewed positively by almost all sectors. However, at the same time, the conservative view has echoed the need for the system sustainability, which may also be arguably right. Both viewpoints have complimentarily and also distinct application values, which need to be accepted. Therefore, one should always look forward to addressing both sustainability and development in Indian agriculture in the years to come.

The changing scenario has recorded agriculture more as a business proposition than merely tradition at least in the agriculturally progressive states. The traditional agriculture, which was a way of life of the majority of Indian farmers, has gradually shrunken to fewer pockets. It is rightly believed that the traditional farming was consistent for cropping patterns, mixed crop composition, cultivation practices, etc., and it was based on the hereditary and the community level experiences. This way of agriculture harboured several friendly and synergistic biota, including, earthworms and soil micro-organisms. The non-biological resources were equally nurtured, for example, the soil structure, composition and organic matter content. In terms of subsistence, the traditional agriculture ensured the farming families more or less self-sufficiency because of an array of harvests even though individual commodities harvested were little in quantities. The farmer did not have to pay for these commodities in the local market. Rather they sometimes earned a little profit by selling their marginal surplus in the weekly haats.

However, the developmental pursuits led by the improved technologies have been affecting the traditional agriculture in terms of both content and extent. This is mainly because of the economic consideration wherein traditional farming has gradually become uneconomic and obsolete. Modern agriculture became commodity or market oriented wherein direct commercial benefit was the main criteria for choice of crop grown by a farmer. The needs of the progressive farmers for commodities other than their main crop(s) that would have been ordinarily met from the traditional mixed farming plots obviously had to be fulfilled from extrinsic sources. Further, in order to organize progressive farming, farmers have to invest in several inputs, such as, tools and farm machinery, improved quality seeds, fertilizers, pesticides, post-harvest processing yards and sheds. As a result, their financial needs for primary investment have been on an increase. This required more production and income. Thus, the economics and financial considerations became increasingly important, including the costs, returns, markets and profits.

The technological break-through in agriculture has also given way to diversification in several ways; with non-traditional crops, new varieties of crops or new methods of cultivation although in terms of nutritional security, the former group is most important. Also, multiple farm enterprises, such as crops, dairying, and poultry, are important from the perspective of both economics and diversification. With the ever-increasing population, rapid urbanization, growing external markets, and more and more demands for farm products competition is also likely to increase. This obviously requires emphasis on fair and efficient regulatory mechanisms and the elaborate legal protection of farmer entrepreneurs and innovators.

In India, crop-livestock mixed farming is traditional. Most of the traditional systems are highly efficient and self-sufficient, and thus are sustainable. The combination of livestock with crop production is an effective risk aversion mechanism, developed out of generations of experience of farmers in arid and rainfed areas. The system is a very good example of recycling of all farming byproducts and household waste with little dependence on outside resources. This is an appropriate and sustainable approach for remote rural areas, where outside resources or services are accessible with great difficulty.

However, there is need to understand production systems and to assess the scope and priorities for development interventions. In semi-arid regions, the combination of trees, cereal crops, leguminous pulses and oilseeds along with a mix of livestock, i.e., cows, buffaloes and goats is common. Each of these is adopted with a multi-purpose objective. Trees provide fruit, fodder and fuel, and some trees improve the soil or serve as windbreak. Crops provide food, fodder and fuel, and leguminous crops improve the soil. Most preferred fodder crops are leguminous species grown in the winter and rotated with cereal crops; thus they provide a much needed protein source for the animals. Livestock are a source of food, fuel, manure, draught power and ready cash in emergency. Thus, these are surely movable assets and investment with social value. Each component has multipurpose value although it is usually assessed singularly.

Similarly, in tribal areas, the combinations generally seen are trees, cereal legume mixed cropping, cows and/or goats and poultry. These farming systems have developed on the perceptions of men and women farmers, tribals and agro pastoralists over the generations. It has been observed that in the semi-arid to arid areas or when the rains fail, income from livestock is the main farm income. In tribal areas, small ruminants replace the livestock, which is semi domesticated rather than being stall-fed. Further, in the better rainfall areas, there is shift towards buffalo. The majority of tribal families and the underprivileged community in general, rear poultry in the traditional backyard system. The main developmental concern in these systems, however, is the verification of improved local technology and its delivery system.

Several increasing problems have been associated with the adoption of new technology to enhance returns from traditional farming. There may be the problems of, for example, soil and water management, choice of crops, technical know-how, pests and diseases, natural hazards, marketing, finance, surplus production, price fluctuations and several more. The challenge is to develop technologies and packages that may improve productivity and production in a sustainable, environment enriching and energy efficient manner but without competing for human food. An important ray of hope, which one can notice in this complex changing scenario of agriculture, is that at least there is a new generation of farmers who are more educated, young and energetic. They could be keen on getting more knowledge about the new technology while carrying forward the family farming traditions more rationally and also logistically.

Emerging Issues and Challenges

Ever increasing human and livestock population particularly in the under developed and the developing world is negating the technology driven progress in agriculture. It is estimated that by 2100 AD, human population will touch the 12 billion mark. Most of the predicted future increase in population will be in the developing countries of Asia, Africa and Latin America. Similarly, in case of India where population is increasing continuously, it is predicted that Indian population will stabilize around 1.4 billion by 2025. The population increased @ 1.9 per cent per annum from 1995 to 2000, and the rate of growth is likely to be around 1.8 and 1.6 per cent per annum during the period 2000-2010 and 2010-2020, respectively. To meet the food demand of the population the yield level is required to be enhanced substantially and also sustainably. This has to be viewed in the present scenario when the emerging issues like, agriculture becoming less remunerative because of fall of prices of major agricultural produce in recent years, surplus production of food grains in last 6-7 years, liberalized economy under post WTO scenario driving towards commercialized agriculture, less attraction of youth to agriculture profession, lack of efforts to agriculture for employment-led growth in rural areas, mismatch between agricultural education and agricultural occupation etc.

Natural resources like soil and water are continuously degrading both in quality and also in quantity. Soils are being constantly degraded and overexploited. Water is another important vital resource for economic development. At the global level, three-fourth of the earth's surface is covered with water and the total water resources amount to 1385.5 million km , comprising 97.3 per cent as salt water and 2.7 per cent as fresh water. Of the latter, 75.2 % occurs as polar ice and glaciers, 22.6 % as ground water, 1.9 % as soil moisture and atmospheric vapour, and 0.3 % in lakes and rivers. This indicates that very limited water is available for domestic, industrial and agricultural use. Fresh water is finite, but renewable through continued hydrologic cycling. India is one of the most well-endowed countries in terms of annual rainfall and has about 4 % of the world's fresh water resources. It is projected that the present per capita per annum water availability of 2001 cubic metres) will reduce to stress level of 1700 cubic metres) in the next 2 to 3 decades. Further the projected reduction in water availability to the agricultural sector from the present share of 89 to about 75 % by 2020 would adversely affect our capacity to produce more food. Future gains in agricultural productivity depend upon integrated development and utilization of surface and ground water resources. The indiscriminate use of canal water is leading to water logging and salinization in the major irrigated commands. The increased ground water extraction has declined water table at an alarming rate, putting an additional burden on farmers in terms of investments, equipments and energy. Further, it is estimated that even after achieving the full irrigation potential, nearly 50 % of the total cultivated area will remain rainfed in the country.

Wide variety of ecological habitats in the world supports an enormous diversity of flora and fauna. Out of 10 hot spots of biodiversity in the world, two are in India in the Western Ghats and North-Eastern region. India has about 8 % of the total existing plants and animals of the world. The factors responsible for erosion of biodiversity are; mono cropped agriculture, introduction of traditional crops to non-traditional areas, introduction of irrigation in desert ecosystems, increased pace of land degradation, such as salinity and indiscriminate use of chemicals. The issues of climate change and global warming caused by the emission of greenhouse gases have emerged as new thrusts. It is projected that South Asia may have an increase in temperature between 0.1 and 0.3°C by 2010, and 0.4 and 2.0°C by 2070. The sea level is expected to rise between 15 and 24 cm over the next century and the lowlying areas may get inundated. Moreover, the ozone depletion may lead to increase in ultraviolet radiation and will result in adverse impact on earth's atmosphere. One of the dominant concerns in the 21st century will be of nutritional security. It is reported that almost half of the pre-school children suffer from moderate to severe malnutrition. In addition, vitamins and micronutrient deficiencies have become widespread. People suffering from these maladies are largely concentrated in rural areas and urban slums. Other than their own labour, they hardly have any resources and skills.

Another challenge before us is to transform the present institutions into new agricultural institutes characterized by different work culture, responsiveness and cost-effectiveness. There is a need to build a knowledge based system. We must use information technology as an instrument of research as well as knowledge sharing. Such efforts would provide us a platform for innovative partnership both nationally and globally. The other major concern is to strengthen our position as producers of public goods in order to serve the poorest of the poor, as well as those who are needy among the developing countries. But the task is complex, especially due to new developments, such as increased involvement of private sector, intellectual property protection and emergence of international regimes including WTO, which may restrict free flow of information and materials. Globalization, as a consequence of WTO, linked with the international trade liberalization, opening up of economies, and a free flow of capital, labour, information and technology is a major paradigm shift making significant changes in the world economy. It has a considerable potential to significantly influence both the food/nutritional security and poverty for better or worse, but its implications and consequences are not yet fully understood. The other concern is to ensure increased and secured funding for Science and Technology. To increase productivity and growth for ensuring better living standards and to reduce poverty and malnutrition, funding support for agricultural research will have to be stepped up to a minimum of 2 per cent of agricultural GDP as is the case with many growing economies of the world.

Future Needs and Approach

Indian agriculture is faced with a great diversity of needs, opportunities and prospects. Most importantly, growth needs to be more rapid, more widely distributed and better targeted in order to reduce the absolute number of people living in poverty and to reduce regional imbalances. In responding to these challenges, more efficient and sustainable use will have to be made of increasingly scarce land, water and germplasm resources. The well endowed irrigated areas which account for 37 per cent of the country's cultivated land currently contribute to about 55 per cent of agricultural production, it is projected that a significant part of the further.

CHAPTER 11

Indian Agricultural Concerns and Future Prospects

In India, economic growth improved significantly during the past two and half decades, particularly in the post-reform period, India is considered as one of the fastest growing economies in the world. However, the exclusion problems have not been seriously addressed by the government programmes and strategies. The experience of the economic reforms during the past 15 years indicates that while there have been improvements in economic growth, foreign exchange, IT revolution, export growth, etc., the income distribution has been unequal and only some sections of the population have benefited more from higher growth and prosperity. One of the excluded sectors during the reform period was agriculture which showed low growth and experienced more farmers' suicides. There are serious concerns on the performance of agriculture sector in the country. The post-reform period growth was led by services. The commodity sector growth (agriculture + industry) has not been higher in the post-reform period as compared to that of 1980s. Particular worry is the agriculture sector which showed lower than 2% per annum growth during the previous decade.

Agriculture not only contributes to overall growth of the economy but also provides employment and food security to majority population which in turn reduces poverty in a developing country. The rapid change in Indian Agriculture particularly since the era of green revolution has been received with mixed response. The development perspective has been definitely encouraging and viewed positively by almost all sectors. The changing scenario has recorded agriculture more as a business proposition than merely tradition at least in the agriculturally progressive states. The traditional agriculture, which was a way of life of the majority of Indian farmers, has gradually shrunken to fewer pockets. It is rightly believed that the traditional farming was consistent for cropping patterns, mixed crop composition, cultivation practices, etc., and it was based on the hereditary and the community level experiences. The experience from BRICS countries indicates that a one percentage growth in agriculture is at least two to three times more effective in reducing poverty than the same growth emanating from non-agriculture sectors.

The Indian Agriculture sector is expected to growth with better momentum in the next few years owing to increase in investment in agricultural infrastructure such as irrigation facilities, warehousing and cold storage. Indian agriculture had growth of 4.6

per cent during 2014 which likely to increase in recent period. India's economic security continues to be depended upon the agriculture sector. In India, agriculture supports 58% of the population, as against about 75% at the time of independence. In the same period, the contribution of agriculture and allied sector to the Gross Domestic Product (GDP) has fallen from 61 to 19%. The 12th Five Year Plan's estimates of expanding the storage capacity to 35 MT and the target of achieving an overall growth of 4 per cent will also go a long way in modifying the overall face of the Indian agriculture sector in the next few years. Further, it may be noted that in the last two Five Year Plans, it is clearly mentioned that for the economy to grow at 9 per cent, it is important that agriculture should grow at least by 4 per cent per annum.

Overall Growth Performance of Agriculture

The agriculture sector in India has undergone significant structural changes in the form of decrease in share of GDP from 30 percent in 1990-91 to 14.5 percent in 2010-11 indicating a shift from the traditional agrarian economy towards a service dominated one. About 52% of the total workforce is still employed by the farm sector which makes more than half of the Indian population dependant on agriculture for sustenance. However, within the rural economy, the share of income from non-farm activities has also increased.

The growth performance of the agriculture sector has been fluctuating across the plan periods. It witnessed a growth rate of 4.8 per cent during the Eighth plan period (1992–97). However, the agrarian situation saw a downturn towards the beginning of the Ninth plan period (1997–2002) and the Tenth plan period (2002–07), when the agricultural growth rate came down to 2.5 percent and 2.4 percent respectively. This crippling growth rate of 2.4 percent in agriculture as against a robust annual average overall growth rate of 7.6 per cent for the economy during the tenth plan period was clearly a cause for concern. The trend rate of growth during the period 1992-93 to 2010-11 is 2.8 percent while the average annual rate of growth in agriculture & allied sectors-GDP during the same period is 3.2 percent.

The increasing divergence between the growth trends of the total economy and that of agriculture & allied sectors suggests an under performance by agriculture. It is also significant that unlike the overall economic growth pattern, agricultural performance in India has been quite volatile. The Indian agriculture growth pattern has been highly varied at the state level. Since agriculture is a state subject, the overall performance of the agriculture sector in India largely depends on what occurs at the state level. There is a wide variation in the performance of different states. During 2000-01 to 2008-09, the growth performance of agriculture in Rajasthan (8.2%), Gujarat (7.7%) and Bihar (7.1%) was much higher than that of Uttar Pradesh (2.3%) and West Bengal (2.4%).

The shares of fruits & vegetables and livestock have shown an increasing trend in recent years implying that they have been growing at a much faster rate than the traditional crops sector. Given the rising share of high value commodities in the total value of agricultural output and their growth potential, this segment is likely to drive agricultural growth in the years to come. Being highly perishable in nature, this segment requires faster and better linkages between farms and firms in terms of logistics, processing and organised retailing. This would entail institutional changes that can incentivise entrepreneurs to invest in building efficient and faster value chains that reduce wastages, and increase the incomes of the farmers at the bottom of the chain.

Future Agriculture and Challenges

1. Fast diminishing per capita availability of soil and water resources and soaring population growth.
2. Decelerating agriculture growth rate, declining total factor productivity low, oscillating and stagnating yield and productivity of major commodities.
3. Huge yield gaps especially in the vast rainfed areas (65% of the total cultivated areas), the widening disparities between the irrigates and rainfed areas and imbalanced agricultural production.
4. Shrinking and degrading production resources, wide spread nutrient deficiencies and toxicities and intensifying biotic and abiotic stresses including climate change and environmental pollution.
5. Declining capital formation and all time low investment in agriculture and burgeoning unemployment.
6. Poor competitiveness of Indian agriculture in the globalised and liberalized market, poor trade and overall agriculture policies, inadequate institutional support and fragile linkages.
7. Imbalanced and eroded agriculture research, education and technology development.
8. Wide regional disparities.
9. Expected climate change and attendant hazards.
10. By 2050, agriculture production must increase by 50% and mostly through yield enhancement and increase in cropping intensity, coupled with sustainability.

Strategies for Future Agriculture

Enhanced Land Productivity

Presently 43% of the geographical areas are used for agricultural activity. It is estimated that out of 329 m ha of geographic area, the potential area for biological production is 265 m ha. If we compare the potential and actual arable land of the contory, the potential arable land is 206.3 m ha and actual arable land is 169.9 m ha, hence, 36.4 m ha difference existing which can be made available for cultivation in future. The pressure of the population on land is increasing, it is 0.18 ha per person at present and estimated to decrease up 0.12 ha per person upto 2050 when the population of country will be stabilize between 1.4 to 1.5 billion. In 2000-01, the percent of area by land holding size was 18.88 below 1 ha, 5189 between 1 to 5 ha and 29.23 per cent a bone 5 ha, however, per cent of farmers by land holding size was 62.81 below 1 ha, 3282 between 1 to 5 ha and 4.3 above 5 ha. Hence, land holding size becoming smaller. It is evident that China with 0.24 ha area per economically active person in agriculture is able to produce double (4,844 kg), cereal grain/ha than India producing (2,232 kg/ha) with 0.63 ha land holding size. Hence, land holding size is neither a resion for lower land productivity nor hinders the enhancement of productivity.

India's global rank in major agricultural crops for area and production ranges between 1 to 7 but for productivity the range is between 7 to 147 which show that potential exists, there is need to bridge the gap. If we compare between 1980-90 and 1990-2000, the area increase in cultivation and productivity was almost half during 1990-2000 than in 1990-2000. From this point, huge and cry started about the stagnation in yield and decline in factor productivity but at the same time there was wide gap between average yield and yield obtained in field demonstration and further the yield obtained at the research station. These situations also indicate that there is passage available to increase the production by bridging the gap. Hence, an increase investment in agriculture sectors will facilitate in expansion of arable land, irrigated area, reducing land and water degradation, undertaking sustainable management of land, water and other natural resources, which will provide the passage to bridge the yield gap with the help of integrated research and technology development and transfer.

Enhanced Water Productivity

Water is one the crucial inputs in agricultural production. Agriculture currently consumes 70 per cent or more of the fresh water supplies. The fact that 2000 to 500 litre of water is required to produce food to feed one person each day, under score the importance of increasing the productivity of water in agriculture. In India, 185 m ha m surface water is available out of which 105 m ha m goes to sea and 10 m ha m is evaporated, only 70 m ha m is available for irrigation (51 m ha m) and for other use (9.0 m ha m). The ground water availability is 40 m ha m out of which 5 m ha m evaporated or use for rising water table only 35 m ha m available for irrigation (26 m ha m) and for other use (9.0 m ha m). Hence, only 77 m ha m water is available for irrigation (surface water 51 m ha m + ground water 26 m ha m).

Water productivity in agriculture varies access regions and crops for example the water productivity in rice cultivation varies from 0.1 to 0.15 kg per cubic meter to 0.32 to 0.41 kg per cubic meter of water in India while in China it varies from 0.4 – 0.6 to 0.95 – 1.71 per cubic meter of water. The water productivity in rice in India is one of the lowest in the world. The declining growth in irrigated area, competing demand for domestic and industrial uses and the fact that 60 per cent of the arable land is still rainfed which calls for efficiency in water use. In the counting water use efficiency varies from 30 to 40 per cent and it was estimated that by reducing the loss, the water use efficiency can be improved upto 10 to 15 per cent. The loss in carrier system is 25 to 40 per cent and due to over irrigation and wastage of irrigation water varies between 45 and 70 per cent. Due to over irrigation or seepage or improper drainage about 8.5 m ha land is affected due to water logging, 9.5 m ha affected due to soil salinity and 9.0 m ha is affected due to flood. Overall, the management of irrigation water in India is one of the poorest in the world. It is predicted that in future the most ferocious competition occur not for oil but for fresh water. The crop wise irrigation availabilities (% of the grass cropped area) is 56.0, 89.5, 21.1, 15.0, 36.1, 92.5 and 28.0 for rice, wheat, maize, pulses, cotton, sugarcane and oilseed, respectively in 2005-06. Higher percentage of water goes to rice, wheat and sugarcane, where irrigation water management is very poor or lot of water goes for wastage. In India, there is wide gap between potential irrigated area created (major, medicor and minor irrigation) and area utilized for irrigation. If we include the cost of water for agricultural produce even at minimal level than even high value crop will also not able to recover only the cost of water. Considering above, the possibility exist for improving the water productivity by improving water use efficiency, reducing the regional variability, reducing the losses with adoption of technology led management like leveling/ smoothing, proper drainage previous, appropriate amount and time of irrigation, water application with inclusion of high tech irrigation system as per their suitability.

Productivity of Other Inputs

Wide disparity in input use access crop is reflected in the variation in productivity differences. The state with highest in the use of inputs are not the ones with higher productivity. Notwithstanding the variation in agro-climate and physical conditions across the state, it is evident that existence of inefficiency in input use is a major cause for low productivity and profitability in agriculture. Fertilizer is the major purchased input which had high yield response in the 70's and 80's. It was one of the important driving force behind green revolution in India. Cereal grain nitrogen use efficiency in word be estimated at 33 per cent, however, use efficiency in developed counting is 42 per cent and in developing counting, it is 29 per cent. At national level, it is estimated that 25.1 mt of nutrient removed from soil and 15.0 mt is supplied from sources, hence, 10.1 mt gap exists. In India, the nitrogen use efficiency ranges between 30 to 40%, P use efficiency between, 15 to 20%, K use efficiency 8 to 10% and micro nutrient use efficiency is less than 10 per cent. The lsos of nutrient through runoff is 1 to 13%, through denitrification 9 to 25% (corn 20%, low land rice 10% and wheat 9.0%), leaching losses 20 to 27%, volatilization losses 7 to 54% and K losses due to surface runoff and leaching is 5 per cent. The P fixation and transformation ranges between 5 to 95 per cent.

At all India level, it was estimated that 55 kg/ha gap exists in fertilizer use. In India 94.8 kg/ha/year fertilizer use recorded but in other country the fertilizer use varies between 150 kg/ha/year to 450 kg/ha/year. It is well established that with increase in fertilizer use, production and productivity improves. In wheat, the three year information w.e.f. 2003-04 to 2005-06 showed that Punjab, Haryana, UP and Bihar uses, 208, 178, 148 and 120 kg NPK/ha, respectively and produces 4209, 3663, 2601 and 1765 kg grain yield per hectare, respectively. It was also noticed that China rank seventh in fertilizer consumption in rice while India rank fourth but rice yield of China is double than the India, it shows the efficiency of applied fertilizer utilization through appropriate technology use. In India, the NPK ratio is recommended 4.00-2.00-1.00 but almost all part of country imbalance. Hence, bridging the gap in fertilizer use, balance use of fertilizer with improving the use efficiency and reducing the losses with technology led management, may substantially improve the production and productivity even at the current level of technology use. Other than the nutrition of the crop, other input like pesticide consumption registered negative growth in last decade. The reversal of trend in pesticide use may be due to popularization of Integrated Pest Management (IPM) strategies, increasing awareness about the health and environmental implication of pesticides, development of resistant cultivars and introduction of more efficient molecules. The consumption of pesticide come down to 275 gram per hectare in 2000-01 from 349 grams in 1977-75. But distribution of certified quality seed increased at a compound growth rate of 5 per cent per annum in 1991-92 to 2001-2002 period.

Climate Change and their Impact

Climate change due to emission of green house gasses has emerged as a new thrust. It is projected that South Asia may have an increase in temperature between 0.1 and 3°C by 2010, and 0.4 to 3.0°C by 2070. At the same time doubling of CO_2 concentration by 2030 with increase average global temperature by 1to3°C. Similarly an increase in the rainfall by 5 to 15 per cent in South Asian region was predicted along with several distressed prediction. The positive change in CO_2 and rainfall can lead to increase crop yield and hence, impact of temperature rise may be neutralized. India as a whole does not have significant impact due to climate change. Even with analyzing the regional variability of climate change the impact can be avoided by changing the land use strategies.

Opportunity/Strategies for Improvement

Identification of sources of inefficiency and the most efficient uses of inputs and available technologies become fundamental to improve agriculture, especially in this competitive globalized world. The approach to raising agricultural productivity should be multi-dimensional. Some of them are mentioned as under.

1. Productivity of input depends upon the level of technology generation and adoption. Many studies supported "fatique' in green revolution technologies. Biotechnology offers powerful new tools for improving agricultural productivity, environmental quality and nutritional quality of staple food. Development of biotechnology as a component of our strategy to increase agricultural productivity and economic growth needs to be explored.
2. The new emerging technologies like super wheat and rice, hybrid wheat and rice, agro-forestry, transgenic crops with resistance to insect and weed will further increase yield and minimize the losses in the field and in storage.
3. The degradation of land and water adversely influencing the productivity. Integrate the management from rivers and reservoirs to rainfall stored in the soil profile with technology and practices that will improve water productivity and reduce degradation of agro-ecosystem.

4. The yield gap (difference in maximum attainable yield with best management practices and the average yield at the farm level) which is very wide (ranging from 14-47% in wheat, 18-70% and 36.77% in rice and in maize) that can be bridged with improved management prentices suited to different ecology. This requires shift in production technologies from farmers resource endowments.
5. Precision farming is an integrated agricultural system incorporating several technologies. At farm level precision farming technology need to be popularized to increase the input use efficiency and to reduce environmental pollution from agro-chemicals.
6. Enhance access to refined and improve the technology consolidating the gain made in the past, extending, the grain in the area deprived of the gains, and enlarging the level of potential of productivity and profitability. The increase emphasis on post harvest management, processing, value addition and marketing through use of frontier technology. This will help close to the realizable yield gap.
7. Pay close attention to input such as irrigation, fertilizer and other agro-chemicals by a system approach such as integrated plant nutrient system, integrated pest management system and integrated farming system.
8. In order to increase agriculture productivity and sustainability, the approaches which will increase water use efficiency, fertilizer and other input use efficiency, increase unskilled labour productivity, agro-forestry, water shed management and farmer managed high-tech-irrigation system, need to exploited.
9. Moreover, strengthen effort to protect the environment, improve commercialization of agriculture, increase investment in human resource, improve trade linkage, improve government policies, knowledge based development, harnessing the youth power, revitalizing the agricultural research and support system and empowering the powerless will certainly remove the imbalances and improve the situation to meet the challenges to feed the increasing population.

Most efficient technology led management will certainly accelerate agricultural development and hence, can easily increase India's food production by 50 per cent in the next 30 years from the current 266 m t to about 400 mt. Which requiring 'proactive' mode of action such as anticipates future demand and develop a road map and action plan, in accordance. In future there is need to develop long term strategic and innovative research initiatives with maximizing the biotech intervention for increased production. Also to develop precision and intensive agricultural technologies for irrigated areas.

The future path of growth in agriculture must be different to ensure congruence among enhanced productivity, sustainability, profitability, equity and ecological security. In addition, resource and situation based harmonization as a changing time scale say increase in input use, larger investment in agriculture section for creation of additional irrigation facilities etc, price and marketing policies keeping in view the need of equity and social justice, would be rewarding. Agricultural production must increase mostly through yield enhancement and increasing in cropping intensity, coupled with sustainability.

These strategies will lead to shift landless workers and marginal farmers from subsistence farming and marginal existence to a better life.

A statutory right to food is both civilized and necessary particularly in present scenario of global food crisis. Ideally the food and nutritional security are the shared responsibility of both society and state. Hence the Public Distribution System (PDS), National Food Security Bill, 2013 as well as Buffer stock are few milestones for ensuring the food security in India, but faulty distribution system, poor implementation of government schemes as well as corruption make the government system defunct. The food production in India has increased from 217 million tonnes in 2006–07 to 265.5 million tonnes in 2013–14 and it was 255.4 million tonnes in 2012–13. Technology Vision 2015 has estimated that India's population would be 1.53 billion in 2035 and will require 392 mt food grains. Similarly, the milk production will be required 230.5 mt compared to present production of 137.7 mt in 2012–13. Again this year, the food production is expected to cross 260 million tonnes due to better climatic conditions. In spite of increase in food grain production, our country has been placed on 67th place out of 88 countries in the world on Global Hunger Index by International Food Policy Research Institute commenting India is among the most food insecure under following indices;

1. 87% of rural population gets less than recommended 2400 calories/day,
2. 64.5% of the urban population gets less than the recommended 2100 calories/day,
3. Over 40% children <6 years are clinically malnourished,
4. 58% of Indian women are anaemic (National Sample Survey),
5. 12 out of 17 states surveyed by state hunger index show alarming levels of hunger, and
6. India's worst performing states have higher levels of hunger than countries such as Zimbabwe and Haiti.

Some of the standards related to food items required and its availability for balanced diet are listed in the table 1 that tell us the story where we stand and what we need. It is a herculean task to feed our burgeoning population which is likely to reach 1.5 billion by 2030 in which we require a minimum of 304 million tonnes of food grains, 175 million tonnes of vegetables, 96 million tonnes of fruits, 170 million tonnes of milk and 21 million tonnes of meat. Balancing the growing food demand with domestic production is unlikely to be as smooth as in the past mainly because of agricultural production systems has come under the confluence of biotic and abiotic stresses, land is limited to be increased under cultivation, the share of water available to agriculture is reducing and also the ground water has already been over exploited in the intensively cultivated north-western India (GOI, 2011) and increasing global warming leading to extreme climatic events like drought, flood, cyclones and heat waves.

Technology has been the key driver for agricultural growth in past with high yielding varieties, fertilizers, conservation technologies, farm mechanization, plant protection etc., but recently it has been observed that the factor productivity of most of the production systems have declined and the crop productivity is not increasing even after applying much more inputs indicating that the these technologies are no more beneficial to increase our food production. Therefore, it is need either to redefine the technologies or look for new technologies that may revolutionize our agricultural production systems.

Conservation technologies including zero or reduced tillage, sub soiling, permanent raised bed planting and laser levelling, water harvesting, watershed management and micro irrigation are the modern technologies that have shown its greater potential to improve crop productivity and farm income with reduced cost of cultivation. Besides, the frontier technologies like biotechnology, nanotechnology, remote sensing and geographical Information System (GIS), information and communication technology (ICT) offer greater opportunities to enhance genetic potential of crops, improve input use efficiency, reduce production and transaction costs and improve sustainability of natural resources.

1. Biotechnology

Biotechnology allows selective genetic modifications of flora and fauna at the genetic molecular level. The specific gene of a unique trait is selected and it is transferred precisely to a desired plant variety by using biotechnological techniques like genetic engineering, genomics, tissue culture, micro propagation and marker assisted breeding. The biotechnology has proved its potential to transform agriculture and agri-food industry in developed countries and so in India it can do it. The biotechnological tools/techniques is to be used to develop new type of seed/plant type that may yield more, resist to many abiotic and biotic stresses. It can modify cereals in to nitrogen fixing crops by transferring the nitrogen fixing gene from legumes. Biotechnology may help to bio-fortified cereals and other crops to improve its nutritive values like development of *golden rice* with high beta carotene content, *high quality protein maize (HQPM)* containing high lysine content, *high starch potato (HSP)* that absorbs less fat during frying and improve human health and nutrition.

The genetically modified crops (GM crops) are developed with the help of biotechnology for specific trait Number of crop varieties of cotton, soybean, canola, maize, papaya, potato, sugar beets and alfalfa have been developed resistant mainly for insect-pest, diseases and weeds and many more yet to be developed for different traits. Transgenic of these crops are available for commercial cultivation. In 2011, biotech crops occupied 160 m ha of land across the globe and half of it was in developing countries (James, 2011). USA is leading in biotech crops with share 43.1%, followed by Brazil 18.9%, Argentina 14.8%, India 6.6%, and China 6.5%. It has also been reported that USA, Brazil, Argentina and Canada occupied genetically modified maize, soybean, canola and sugar beet around 90% of the area under their cultivation. Brookes and Barfoot (2012) reported that the biotech crops had added addition US$78 billion to the global economy by way of improved yields and reduced input costs. In addition, the biotech crops during the same period produced 276 million tonnes of food, feed and fibre equivalent to 91 million hectares land and also reduced pesticide use by 443 million kg (active gradient).

The success story of Bt cotton is well known in the world. In India it was introduced in 2002 for control of mainly cotton boll worm (*Helicoverpa armigera*) and since then the area under Bt cotton has increased to 90% of the total area in 2011. It has transformed Indian cotton industry with 3.5fold increase in cotton production and substantial rise in export from 0.05 billion bales in 2001-02 to 8.3 billion bales in 2009-10. Indian farmers have reported the yield increase up to 30%, 40% reduction in insecticide usage and 50% more farm income. Besides employment generation due to bumper cotton production, income opportunities in non-farm business like trading activities and saving of other crops from the same polyphagus insect are other benefits of Bt cotton. Similarly Bt Brinjal has been developed in India against the shoot and fruit borer and it can reduce insecticide usage by 42-77% and enhance yield by 37-55% (Krishna and Quim, 2008 and Kumar et al., 2010). Similarly, the crop varieties resistant to stresses like drought, water logged, saline & alkaline soils, cold etc may be developed with the help of biotechnological tools. The benefit of C_4 plants may be transferred to C_3 plants. Presently the research on breeding for delayed ripening of fruits and vegetables is underway. James (2010) reported that presently 11 crops i.e. brinjal, cotton, groundnut, mustard, papaya, potato, rice, sorghum, sugarcane, tomato and watermelon under public and 8 crops i.e. brinjal, cabbage, cauliflower, cotton, maize, okra, rice, and tomato under private sectors are in the process of genetic modification.

The biotechnology is also important to contribute towards improving the quality of natural resources. Microbial biotechnology based products like bio-fertilizers and bio-pesticides improve soil health and preserve the environment. Similarly the micro-organisms have varieties and have to explore for the welfare of human beings particularly in bio-degradation of vast household and agricultural wastes, preserve food products, increase food quality, improve degraded soil etc. In livestock development too, biotechnology may help in many ways like artificial insemination, sexed semen technology, genetically modified cows and other livestock, fermentation and microbial technologies, vaccines, disease diagnosis and food preservation.

2. Nanotechnology

In modern era we need precision in every walk of life so in agriculture. There are wide variations in recommended agro techniques for different crops but these techniques are most often generalized. For example, the recommended wheat seed rate is 100 kg/ha and fertilizer dose is 120: 60:40::N:P_20_5:K_20 kg/ha in most of the literature without consideration of early, normal and late sown wheat, therefore this type of flat recommendations are faulty and need precision recommendations. Precision agriculture not only save energy and cost but also improve yield with sustained soil quality. Nanotechnology is the new technology that may contribute to precision agriculture in many ways like controlled release of fertilizers, stabilization of bio pesticides, slow release of nano material assisted fertilizers, bio-fertilizers and micronutrients for efficient use and nano material assisted delivery of genetic material for crop improvement (Ghormade *et al.*, 2011). Nano sensors can be used for detecting pathogens and pesticides, and for soil conservation and remediation, and also help to improve input use efficiency, reduce input losses and assisting in the precision farming. Nanotechnology is also useful in protecting the environment through use of alternative energy supplies, and filter and catalysts to reduce pollution and clean up existing pollutants. The potential applications of nanotechnology in food industry including packaging, storage and processing of food commodities for improving quality safety and self life (Sastry et al., 2011), especially of perishable commodities that suffer heavy post harvest losses.

3. Remote Sensing and GIS Application

Remote sensing, geographical information system (GIS) and global positioning system (GPS) are the modern tools that are used to collect earth information from satellites, check its truthfulness with the help of GPS and then analyzed with the help of GIS and finally the decisions are drawn in very short time, hence the earth information of large area is quickly collected and analyzed for decisions and its implementation quickly. Therefore RS and GIS have a great potential for precision agriculture. Presently we have a number of satellites in space that can draw earth information of < 1 m accuracy; therefore these remotely sensed information is very useful in agricultural development. The information on crop area, water bodies, rainfed and irrigated area, soil erosion, ground water scenario, crop growth stages and its conditions, incidence of pest and diseases, advance crop yield prediction, flood, tsunami, coastal tides, salinity and alkalinity, desertification etc. is collected and may be utilized by many stakeholders like farmers, planners, and public and private institutions for hi-tech agriculture for sustaining our agriculture and livelihood in future.

4. Information and Communication Technology (ICT)

India is blessed with enormous bio-resources that include a wide variety of crops and a favourable environment to utilize this abundance and genetic diversity. India is divided into 16 agro-climatic zones, 10 vegetative zones and 15 biotic provinces. Rice cultivation in India covers nearly 22% of agricultural area, which is followed by wheat (13%), oilseeds (15%), pulses (13%), commercial crops (8 %) and horticultural crops (29%). The majority of Indian agriculture consists of farms with land holdings of less than one hectare (59%) and farm with holdings of one to two hectares (19%). This represents nearly 32 % of the total area under cultivation. The average per capita availability of land has decreased to 0.15 ha. Nearly 50 % of the arable area is rainfed although there has been a significant increase in irrigation acreage (80.6 mha).

The role of Information Technology to develop agricultural research, education and extension to improve quality of life in rural area is well established. IT can help an average Indian farmer to get relevant information regarding agro-inputs, crop production technologies, agro processing, market support, agro-finance and management of farm agri-business. The agricultural extension mechanism is becoming dependent on IT to provide appropriate and location specific technologies for the farmers to furnish timely and proficient advice to the farmers IT can be a best mean not only to develop agricultural extension but also to expand agriculture research and education system. The excellent scopes of Internet are waiting to develop agricultural education management through smart exposure of agricultural teachers and educational planners, class rooms, virtual class as well as dropout agricultural learners. To prepare the agricultural graduates capable to meet the challenges of the new millennium, they should be given course on International Agriculture, WTO, Trade Related Intellectual Property Rights (TRIPs), Global Conventions on Climate, Biodiversity and Desertification, Computer Technology, Patent and Trade Literacy, International Standards. For this IT can play important role. So Introduction to Computers, Application of Software, Data Base Management Systems, PowerPoint, Drawing Software, Computer Programming, Multimedia, Internet, role of TV and Radio in ICT should be a part of their course curriculum. For agricultural extension management, role of IT can be encouraged for future resource documentation, as methods of extension and linkage between research and extension. IT in agricultural research management for textual and non textual documentations and deciding prioritization of research areas needs to be reinforced. The crop forecasting, input management, command area management, watershed management, land and water resources development, drinking water potential mapping precision management, natural disaster management, fishery management, hill area development and post harvest management are the key areas; where Information Technology can play its imperative impact.

CHAPTER 12

Indian Council of Agriculture Research (ICAR) Institutes

The ICAR and the national agricultural research system have played a pioneering role in developments in agriculture in India through its research and technology development that has enabled the country to increase the production of food grains by 4 times, horticultural crops by 6 times, fish by 9 times (marine 5 times and inland 17 times), milk 6 times and egg 27 times since 1950-51, thus making a visible impact on the national food and nutritional security. It has played a major role in promoting excellence in higher education in agriculture. It is engaged in cutting edge areas of science and technology in their fields. India has one of the largest agricultural research systems in the world with the largest number of scientific personnel of any developing country engaged in research and education relating to agriculture and allied areas.

The ICAR is an autonomous organization under the Department of Agricultural Research and Education (DARE), Ministry of Agriculture, Govt. of India. Formerly known as Imperial Council of Agricultural Research, it was established on 16th July, 1929 as a registered society under the Societies Registration Act, 1860, in pursuance of the report of the Royal Commission on Agriculture. The Council is the apex body for coordinating, guiding and managing research and education in agriculture including horticulture, fisheries and animal sciences in the entire country. With 102 ICAR institutes (including four deemed universities) and 65 Agricultural Universities spread across the country this is one of the largest national agricultural system in the world.

Research Infrastructure of ICAR

ICAR has established many Central Research Institutions over the years to meet the agricultural research needs of the country. They are given below-

- Deemed Universities-4
- ICAR Institutions-65
- National Bureaux-6
- Project Directorates-13
- National Research Centres-14
- Substations of ICAR Institutes-138

- AICRPs -61
- Other Projects-10
- Network Projects-19
- Zonal Project Directorates-8
- State Agricultural Universities (SAUs)-64
- Central Agricultural University -3
- Central Universities having faculty of agriculture-4
- Krishi Vigyan Kendras (KVKs) - 706

Deemed Universities

1. Indian Agricultural Research Institute, New Delhi
2. Indian Veterinary Research Institute, Izatnagar
3. National Dairy Research Institute, Karnal, Haryana
4. Central Institute of Fisheries Education, Mumbai

ICAR Institutes

1. ICAR-Central Island Agricultural Research Institute , Port Blair
2. ICAR-Central Arid Zone Research Institute, Jodhpur
3. ICAR-Central Avian Research Institute, Izatnagar
4. ICAR-Central Inland Fisheries Research Institute, Barrackpore
5. ICAR-Central Institute Brackishwater Aquaculture, Chennai
6. ICAR-Central Institute for Research on Buffaloes, Hissar
7. ICAR-Central Institute for Research on Goats, Makhdoom
8. ICAR-Central Institute of Agricultural Engineering, Bhopal
9. ICAR-Central Institute for Arid Horticulture, Bikaner
10. ICAR-Central Institute of Cotton Research, Nagpur
11. ICAR-Central Institute of Fisheries Technology, Cochin
12. ICAR-Central Institute of Freshwater Aquaculture, Bhubneshwar
13. ICAR-Central Institute of Research on Cotton Technology, Mumbai
14. ICAR-Central Institute of Sub Tropical Horticulture, Lucknow
15. ICAR-Central Institute of Temperate Horticulture, Srinagar
16. ICAR-Central Institute on Post harvest Engineering and Technology, Ludhiana
17. ICAR-Central Marine Fisheries Research Institute, Kochi
18. ICAR-Central Plantation Crops Research Institute, Kasargod
19. ICAR-Central Potato Research Institute, Shimla
20. ICAR-Central Research Institute for Jute and Allied Fibres, Barrackpore

21. ICAR-Central Research Institute of Dryland Agriculture, Hyderabad
22. ICAR-National Rice Research Institute, Cuttack
23. ICAR-Central Sheep and Wool Research Institute, Avikanagar, Rajasthan
24. ICAR- Indian Institute of Soil and Water Conservation, Dehradun
25. ICAR-Central Soil Salinity Research Institute, Karnal
26. ICAR-Central Tobacco Research Institute, Rajahmundry
27. ICAR-Central Tuber Crops Research Institute, Trivandrum
28. ICAR-ICAR Research Complex for Eastern Region, Patna
29. ICAR-ICAR Research Complex for NEH Region, Barapani
30. ICAR-Central Coastal Agricultural Research Institute, Ela, Old Goa, Goa
31. ICAR-Indian Agricultural Statistics Research Institute, New Delhi
32. ICAR-Indian Grassland and Fodder Research Institute, Jhansi
33. ICAR-Indian Institute of Agricultural Biotechnology, Ranchi
34. ICAR-Indian Institute of Horticultural Research, Bengaluru
35. ICAR-Indian Institute of Natural Resins and Gums, Ranchi
36. ICAR-Indian Institute of Pulses Research, Kanpur
37. ICAR-Indian Institute of Soil Sciences, Bhopal
38. ICAR-Indian Institute of Spices Research, Calicut
39. ICAR-Indian Institute of Sugarcane Research, Lucknow
40. ICAR-Indian Institute of Vegetable Research, Varanasi
41. ICAR-National Academy of Agricultural Research & Management, Hyderabad
42. ICAR-National Institute of Biotic Stresses Management, Raipur
43. ICAR-National Institue of Abiotic Stress Management, Malegaon, Maharashtra
44. ICAR-National Institute of Animal Nutrition and Physiology, Bengaluru
45. ICAR-ICAR-National Institute of Natural Fibre Engineering and Technology, Kolkata, Kolkata
46. ICAR-National Institute of Veterinary Epidemiology and Disease Informatics, Hebbal, Bengaluru
47. ICAR-Sugarcane Breeding Institute, Coimbatore
48. ICAR-Vivekananda Parvatiya Krishi Anusandhan Sansthan, Almora
49. ICAR-Central Institute for Research on Cattle, Meerut, Uttar Pradesh
50. ICAR-National Institute of High Security Animal Diseases, Bhopal
51. ICAR-Indian Institute of Maize Research,New Delhi
52. ICAR- Central Agroforestry Research Institute , Jhansi
53. ICAR-National Institute of Agricultural Economics and Policy Research, New Delhi

54. ICAR- Indian Institute of Wheat and Barley Research, Karnal
55. ICAR- Indian Institute of Farming Systems Research, Modipuram
56. ICAR- Indian Institute of Millets Research, Hyderabad
57. ICAR- Indian Institute of Oilseeds Research, Hyderabad
58. ICAR- Indian Institute of Oil Palm Research, Pedavegi, West Godawari
59. ICAR- Indian Institute of Water Management, Bhubaneshwar
60. ICAR-Indian Institute of Rice Research, Hyderabad
61. ICAR- Central Institute for Women in Agriculture, Bhubaneshwar
62. ICAR-Central Citrus Research Institute, Nagpur
63. ICAR-Indian Institute of Seed Research, Mau
64. ICAR-Indian Agricultural Research Institute, Post Box No. 48, Hazaribag 825 301, Jharkhand
65. ICAR-National Institute for Plant Biotechnology, New Delhi

Bureaus

1. National Bureau of Agricultural Insect Resources, Bengaluru
2. National Bureau of Agriculturally Important Micro Organisms, Mau Nath Bhanjan, Uttar Pradesh
3. National Bureau of Animal Genetic Resources, Karnal, Haryana
4. National Bureau of Fish Genetic Resources, Lucknow, Uttar Pradesh
5. National Bureau of Plant Genetic Resources, New Delhi
6. National Bureau of Soil Survey & Land Use Planning, Nagpur, Maharashtra

Project Directorates

1. ICAR - Directorate of Poultry Research, Hyderabad
2. ICAR - Directorate of Groundnut Research, Junagadh, Gujarat
3. ICAR - Directorate of Cashew Research, Puttur
4. ICAR - Directorate of Cold Water Fisheries Research, Bhimtal, Nainital
5. ICAR - Directorate of Floriculture Research, Pune, Maharashtra
6. ICAR - Directorate of Knowledge Management in Agriculture (DKMA), New Delhi
7. ICAR - Directorate of Medicinal and Aromatic Plants Research, Anand
8. ICAR - Directorate of Mushroom Research, Solan
9. ICAR - Directorate on Onion and Garlic Research, Pune
10. ICAR - Directorate of Rapeseed & Mustard Research, Bharatpur
11. ICAR - Directorate of Seed Research, Mau

12. ICAR - Directorate of Soybean Research, Indore
13. ICAR - Directorate of Water Management Research, Bhubaneshwar
14. ICAR - Directorate of Weed Science Research, Jabalpur
15. ICAR - Project Directorate on Foot & Mouth Disease, Mukteshwar

National Research Centers

1. ICAR - National Research Centre on Meat, Hyderabad
2. ICAR - National Centre for Agril. Economics & Policy Research, New Delhi
3. ICAR - National Research Centre on Pig, Guwahati
4. ICAR - National Research Centre for Agroforestry, Jhansi
5. ICAR - National Research Centre for Banana, Trichi
6. ICAR - National Research Centre on Camel, Bikaner
7. ICAR - National Research Centre for Makhana(foxnut), Darbhanga
8. ICAR - National Research Centre on Equines, Hisar
9. ICAR - National Research Centre for Citrus, Nagpur
10. ICAR - National Research Centre for Grapes, Pune
11. ICAR - National Research Centre for Litchi, Muzaffarpur
12. ICAR - National Research Centre for Pomegranate, Solapur
13. ICAR - National Research Centre on Mithun, Medziphema, Nagaland
14. ICAR - National Research Centre on Orchids, Pakyong, Sikkim
15. ICAR - National Centre for Integrated Pest Management, New Delhi
16. ICAR - National Research Centre on Plant Biotechnology, New Delhi
17. ICAR - National Research Centre on Seed Spices, Ajmer
18. ICAR - National Research Centre on Yak, Dirang, Arunachal Pradesh

Agriculture Technology Application Research Institute (ATARIs)

There are eight ATARIs located at Ludhiana, Kolkata, Umiam, Kanpur, Hyderabad, Jodhpur, Jabalpur and Bangalore. In addition to its institute – based research, ICAR promotes research schemes / projects in agriculture and allied activities to resolve location-specific problems such as All India Coordinated Research Projects (AICRPs), Network projects and other projects. This promotes a coordinated and a cooperative endeavor with other research organizations in carrying out multidisciplinary research programmes.

All India Coordinated Research Projects-59

1. AICRP on Nematodes, New Delhi
2. AICRP on Maize, New Delhi
3. AICRP Rice, Hyderabad

4. AICRP on Chickpea, Kanpur
5. AICRP on MULLARP, Kanpur
6. AICRP on Pigeon Pea, Kanpur
7. AICRP on Arid Legumes, Kanpur
8. AICRP on Wheat & Barley Improvement Project, Karnal
9. AICRP Sorghum, Hyderabad
10. AICRP on Pearl Millets, Jodhpur
11. AICRP on Small Millets, Bangalore
12. AICRP on Sugarcane, Lucknow
13. AICRP on Cotton, Coimbatore
14. AICRP on Groundnut, Junagarh
15. AICRP on Soybean, Indore
16. AICRP on Rapeseed & Mustard, Bharatpur
17. AICRP on Sunflower, Safflower, Castor, Hyderabad
18. AICRP on Linseed, Kanpur
19. AICRP on Sesame and Niger, Jabalpur
20. AICRP on IPM and Biocontrol, Bangalore
21. AICRP on Honey Bee Research & Training, Hisar
22. AICRP -NSP(Crops), Mau
23. AICRP on Forage Crops, Jhansi
24. AICRP on Fruits, Bangaluru
25. AICRP Arid Zone Fruits, Bikaner
26. AICRP Mushroom, Solan
27. AICRP Vegetables including NSP vegetable, Varanasi
28. AICRP Potato, Shimla
29. AICRP Tuber Crops, Thiruvananthapuram
30. AICRP Palms, Kasaragod
31. AICRP Cashew, Puttur
32. AICRP Spices, Calicut
33. AICRP on Medicinal and Aromatic Plants including Betelvine, Anand
34. AICRP on Floriculture, New Delhi
35. AICRP in Micro Secondary & Pollutant Elements in Soils and Plants, Bhopal
36. lAICRP on Soil Test with Crop Response, Bhopal
37. AICRP on Long Term Fertilizer Experiments, Bhopal
38. AICRP on Salt Affected Soils & Use of Saline Water in Agriculture, Karnal

39. AICRP on Water Management Research, Bhubaneshwar
40. AICRP on Ground Water Utilisation, Bhubaneshwar
41. AICRP Dryland Agriculture, Hyderabad
42. AICRP on Agrometeorology, Hyderabad including Network on Impact adaptation & Vulnerability of Indian Agri. to Climate Change
43. AICRP Integrated Farming System Research, Modipuram including Network Organic Farming
44. AICRP Weed Control, Jabalpur
45. AICRP on Agroforestry, Jhansi
46. AICRP on Farm Implements & Machinery, Bhopal
47. All India Coordinated Research Project on Ergonomics and Safety in Agriculture
48. AICRP on Energy in Agriculture and Agro Based Indus.,Bhopal
49. AICRP on Utilization of Animal Energy (UAE), Bhopal
50. AICRP on Plasticulture Engineering and Technologies, Ludhiana
51. AICRP on PHT, Ludhiana
52. AICRP on Goat Improvement, Mathura
53. AICRP- Improvement of Feed Sources & Nutrient Utilisation for raising animal production, Bangalore
54. AICRP on Cattle Research, Meerut
55. AICRP on Poultry, Hyderabad
56. AICRP-Pig, Izzatnagar
57. AICRP Foot and Mouth Disease, Mukteshwar
58. AICRP ADMAS, Bangalore
59. AICRP on Home Science, Bhubaneshwar

Network Projects-20

1. All India Network Project on Pesticides Residues, New Delhi
2. All India Network Project on Underutilised Crops, New Delhi
3. All India Network Project on Tobacco, Rajahmundry
4. All India Network Project on Soil Arthropod Pests, Durgapura
5. Network on Agricultural Acarology, Bangalore
6. Network on Economic Ornithology, Hyderabad
7. All India Network Project on Rodent Control, Jodhpur
8. All India Network Project on Jute and Allied Fibres, Barrackpore
9. Network project on Improvement of Onion & Garlic, Pune
10. Network Bio-fertilizers, Bhopal

11. Network Project on Harvest & Post Harvest and Value Addition to Natural Resins & Gums, Ranchi
12. Network project on Animal Genetic Resources, Karnal
13. Network Project on R&D Support for Process Upgradation of Indigenous Milk products for industrial application Karnal
14. Network Programme on Sheep Improvement, Avikanagar
15. Network Project on Buffaloes Improvement, Hisar
16. Network on Gastro Intestinal Parasitism, Izatnagar
17. Network on Haemorrhagic Septicaemia, Izatnagar
18. Network Programme Blue Tongue Disease, Izatnagar
19. Network Project on Conservation of Lac Insect Genetic Resources, Ranchi
20. Network Project on Agricultural Bioinformatics and Computational Biology, New Delhi

Other Projects-10

1. Technology Mission on Cotton (CICR, Nagpur)
2. Technology Mission on Jute (CRIJAF, Barrackpore)
3. Continuation, Strengthening and Establishment of Krishi Vigyan Kendras
4. Strengthening & Development of Higher Agricultural Education in India, New Delhi
5. Central Agricultutral University, Imphal
6. Strengthening and Modernization of ICAR Headquarters
7. Intellectual Property Management & Transfer/Commercialisation of Agricultural Technology (Upscaling of existing component IPR HQ)
8. Indo US Knowledge Initiative
9. National Agricultural Innovative Project, New Delhi
10. National Fund for Basic and Strategic Research, New Delhi

Some Model Questions for Agricultural Heritage

Fill in the blanks with appropriate word: **[0.5 × 8 = 4.0]**

1. __________ is considered to be the birth place of Agricultural Revolution in Western India.
2. It is estimated that more than __________ species of marine organisms may be available in the seas of the world, of which __________ has been identified and in India so far __________ has been identified.
3. The average monthly per capita consumer expenditure in 2007-08 stood at Rs __________ in urban and Rs __________ in rural India.
4. In the year of __________ B.C., cultivation of rice started.
5. During Kanuj Empire of Harsha (606-607 A.D.), a rice variety of large grains of extraordinary fragrance called __________ was grown in Magadh.
6. With reference to the 1950, the productivity gains are nearly __________ times in foodgrains, __________ times in fruits, __________ times in vegetables, __________ times in fish, and __________ times in egg production at present.
7. Geographical area of India is __________, cultivated area of India is __________ and cropping intensity is about __________
8. Buddhist period (600 B.C.) marks the importance of trees. It can be called as a period of __________ and __________
9. Agricultural colleges were established in India during the period of __________.
10. The full form of A.D. is __________.
11. Indian agriculture began by __________ BCE as a result of early cultivation of plants and domestication of crops and animals.
12. *Ayurveda* was mentioned in __________ period.
13. Storage rooms may be smoked regularly with __________ leaves to keep away insects.
14. Plough was invented in __________ BC.
15. Agriculture is __________ years old.

16. __________ crop was introduced by Spaniards in India.
17. __________ oilseed crop was introduced by Portuguese in India.
18. __________ cereal crop was domesticated in India.
19. Maize seeds are soaked in __________ for 10–12 hours before sowing to increase the resistance against insect pest as ITK.
20. The __________ period is known as the early Food Producing era.
21. During Gupta period, cultivators were called by various terms as __________.
22. Irrigation was developed during __________ civilization.
23. __________ storage system designed in early time is also used in present day Agriculture.
24. __________ was one of commercial crops grown during British Raj in India.
25. Agricultural colleges were established in India during the period of __________.
26. __________ rotation was practiced during Sangam period.
27. Fertilizer/pesticide in Agriculture was introduced in __________ century.
28. The full form of A.D. is __________.
29. Indian agriculture began by __________ BCE as a result of early cultivation of plants and domestication of crops and animals.
30. Storage rooms may be smoked regularly with __________ leaves to keep away moths, weevils and beetles.
31. Rice cultivation was started in __________ BC.
32. American Boll worm of cotton can be controlled by the plant extract (biopesticide) of __________ crop.
33. Men concentrated on __________ and __________ during Paleolithic and Neolithic periods.
34. Shifting cultivation is __________ and __________ method of cultivation.
35. Women were the pioneers for cultivating __________ from the wild flora.
36. Maize cultivation was started in __________ BC.

Match the following questions: **[0.25 × 8 = 2.0]**

Column A	Column B	Answer
1. Department of Agriculture	(a) 1972	
2. Imperial Agricultural Research Institute	(b) 1929	
3. NABARD	(c) 1957	
4. First Agricultural University at Pantnagar	(d) 1905	
5. ICRISAT	(e) 1942	
6. Imperial Council of Agricultural Research	(f) 1871	
7. All India Maize Improvement Project	(g) 1960	
8. Department of Food	(h) 1982	

Match the following questions: [0.5 × 10 = 5.0]

Column A	Column B	Answer
1. NABARD	(a) 1972	
2. First Agricultural University at Pantnagar	(b) 2015	
3. ICRISAT	(c) 1957	
4. All India Co-ordinated Research Project on Maize	(d) 1963	
5. Department of Food	(e) 1942	
6. Training & Visit system	(f) 2004	
7. Soil Health Card Scheme	(g) 1960	
8. National Seed Corporation	(h) 1982	
9. Kisan Call Centre	(i) 2007	
10. RKVY	(j) 1977	

Column A	Column B
1. NABARD	(i) 1972
2. First Agricultural University at Pantnagar	(j) 1880
3. ICRISAT	(k) 1957
4. All India Maize Improvement Project	(l) 1963
5. Department of Food	(m) 1942
6. Training & Visit system	(n) 1976
7. Famine commission appointed	(o) 1960
8. NSC	(p) 1982
9. IRDP	(q) 1965
10. IAAP	(r) 1977

Write the full form of the following word:

1. NABARD
2. RKVY
3. FRI
4. AICRP
5. IADP
6. DPAP
7. NAEP
8. ATMA
9. NMSA

10. CRRI
11. NAIP
12. CAD
13. NSC

C. Put '√' mark on appropriate answer of the following questions: [0.5 × 6 = 3.0]

1. Wheat and barley were domesticated in the Indian sub-continent in.
   (a) 9000 BCE (b) 1500 BCE (c) 1500 CE (d) 8000 BCE
2. Archeological evidence of an animal-drawn plough in the Indus Valley civilization was found in
   (a) 2500 BCE (b) 4500 BCE (c) 5440 BCE (d) 3200 BCE
3. Megasthenes was
   (a) Greek diplomat (b) American diplomat
   (c) Latin diplomat (d) Portuguese diplomat
4. Which of the followings is not originated in India
   (a) Cotton (b) Citric fruits (c) Sugarcane (d) Maize
5. In 1966, semi-dwarf rice variety, IR 8 was introduced from and 's respectively is formed as base for green revolution.
   (a) Taiwan (b) Philippines (c) Mexico (d) Malaysia
6. Litchi crop was introduced in India from.
   (a) China (b) Spain (c) Mexico (d) Latin America
7. For controlling of termites, common salt was used as ITK in ________ crop.
   (a) Maize (b) Sugarcane (c) Chilli (d) Paddy
8. Bacterial blight was controlled through ITK by using
   (a) Leaf extracts of Neem and Tobacco (b) cattle urine
   (c) None of these (d) All of these
9. In 1966, semi-dwarf rice variety, IR 8 was introduced from and s respectively is formed as base for green revolution.
   (a) Taiwan (b) Philippines (c) Mexico (d) Malaysia

Answer the following questions: [1.0 × = .0]

(i) Mention the different periods of Agricultural Development before India's independence.
(ii) Write only the salient features of ITK.
(iii) Define ITK.
(iv) How Amrit Pani was prepared in ancient time?
(v) What are the Vavilov's centre of origin of crops?

(vi) How the rain bird concept was used to predict the rainfall in ancient time?

(vii) Mention the different periods of Agricultural Development before India's independence.

(viii) Define Agricultural Heritage.

(ix) What do you mean by Biopesticide?

(x) Describe the use of lemon grass as biopesticide.

(xi) What is Agricultural Heritage?

(xii) How many KVKs and NRc are established in India at present?

(xiii) Mandate of ICAR

(xiv) Current scenario of Indian Agriculture

(xv) Ancient Agricultural practices

(xvi) History of Agriculture during early history period

(xvii) Crop classification

(xviii) Agricultural resources of India

(xix) Features of Neolithic culture in India

(xx) Write in short about national agricultural system in India.

(xxi) Describe in brief about the history of agriculture during pre-independence.

(xxii) Describe the present status of farmers in India.

(xxiii) Mention in details the scope of agriculture in India.

Answer the following questions as 'True' or 'False':

1. During Bronze Age, the early Aryans lived.
2. In the Vedic age, farming was practiced by the people having wisdom only.
3. Livestock farming was initiated in the 18th Century.
4. Buddhist period was in Seventh century BC.
5. Department of Agriculture was created in India during British Rule.
6. During Ashoka period, forestry and horticulture were not promoted.
7. During the Muslim rule, land revenue system was not improved in India.
8. ITK stands for Indigenous Technical Knowledge.
9. Green Revolution refers to increase in production of milk.
10. Imperial (now Indian) Council of Agricultural Research was established in 1929.

Some Model Questions for JRF General Part I

Fill in the blanks with app

1. The organic matter content of soil is obtainable if its organic carbon content is multiplied by the factor of
   (a) 1.72 (b) 1.29 (c) 1.43 (d) 1.89
2. The concept of crop logging was first used in
   (a) USA (b) Cuba (c) Hawaii (d) Brazil
3. The country having 100% cropped area under irrigation is
   (a) USA (b) Egypt (c) Canada (d) UK
4. Rothamstand Agril. Experimental Station in England was founded in the year
   (a) 1865 (b) 1942 (c) 1843 (d) 1924
5. Sunflower is used as indicator plant for
   (a) FC (b) Hygroscopic coefficient
   (c) PWP (d) None
6. The N percentage in wheat protein is
   (a) 10% (b) 6% (c) 11% (d) 16%
7. The term potential salinity was introduced by
   (a) Doneen (b) I. C. Gupta (c) Joseph Fourier (d) Eaton
8. Most critical stage for irrigation in maize is
   (a) Knee height (b) Tasseling
   (c) Grain development(d) Silking
9. Stevenson screens door should face
   (a) East (b) North (c) South (d) West
10. Napier grass will remain dormant in the month of
    (a) June - July (b) February - March
    (c) November - December (d) None

11. Zn chelate is
    (a) Soluble and quick acting (b) Soluble but slow acting
    (c) Insoluble and slow acting (d) Insoluble but quick acting
12. How many pounds of NH_4NO_3 (34% N) would be required to supply 150 kg of N
    (a) 150 (b) 44 (c) 441 (d) 241
13. The colour of land capability class - I on LCC maps is
    (a) Yellow (b) Pink (c) Green (d) Red
14. Moor soils are
    (a) Normal soils (b) Transitional soils
    (c) Abnormal soils (d) None
15. is an objection weed in methi
    (a) *Amaranthus sp* (b) *Melilotus alba* (c) *Echinochloa sp* (d) *Agropyron repens*
16. Brace roots are found in
    (a) Wheat (b) Maize (c) Rice (d) Barley
17. *Avena sativa* is
    (a) Diploid (b) Tetraploid (c) Hexaploid (d) Haploid
18. The optimum plant population for wheats around
    (a) 1-2 lacs (b) 2-3 lacs (c) 3-4 lacs (d) 4-5 lacs
19. The inflorescence of jute is
    (a) Raceme (b) Cymose (c) Siliqua (d) Capsule
20. Hot water treatment of sugarcane setts is done because
    (a) It increases germination
    (b) It controls the soil borne diseases
    (c) It breaks dormancy
    (d) It promotes the conversion of sucrose to glucose
21. Species of wheat not cultivated at present in India is
    (a) *Triticum aestivum* (b) *Triticum diococcum*
    (c) *Triticum durum* (d) *Triticum sphaerococcum*
22. Kharif, Rabi and Zaid words belong to
    (a) Urdu (b) Latin (c) Arabic (d) Greek
23. Natural Farming was advocated by
    (a) Gilbert (b) Fakuoka
    (c) Lawes (d) Norman E. Borlaug

24. __________ *element is essential for sugar translocation in sugarcane.*

(a) K (b) Mg (c) P (d) N

25. Rice protein is located at ____________- position in the grain.

(a) Endosperm (b) Embryo (c) Whole grain (d) Aleurone layer

26. AICRP, Agrometeorrology situated at

(a) Pune (b) Hydrabad (c) Mysore (d) Kochi

27. NIASM situated at

(a) Hydrabad (b) Kochi (c) New Delhi (d) Malegaon

28. The ex-officio President of the ICAR Society is

(a) Prime Minister

(b) President of India

(c) Union Minister of Agriculture for state affairs

(d) Union Minister of Agriculture

29. Keshar(sefforon) belongs family-

(a) Composite (b) Leguminosae (c) Iridaceae (d) Euphorbiaceae

30. Number of insecticide included in insecticide schedule

(a) 776 (b) 756 (c) 787 (d) 780

31. Pure culture techniques for isolation of micro-organisms was developed by

(a) Brefeld (b) Iwanowsky (c) Petri (d) Edward Jenner

32. Total carbon fixed by land per year =

(a) 273×10^{11} (b) 110×10^{12} (c) 110×10^{11} (d) 273×10^{12}

33. Which one is Anti-auxin

(a) NAA (b) 2, 4 –D

(c) MCPA (d) Naphthythalamic acid

34. Rural Credit Review Committee was constituted in

(a) 1998 (b) 1983 (c) 1974 (d) 1960

35. Micro propagation technique was first devised for

(a) Rice (b) Vegetables (c) Orchids (d) Banana

Some Model Questions for JRF General Part II

1. The relationship of climatic regimes and agricultural production is termed as
   (a) Climatology (b) Agricultural meteorology
   (c) Agroclimatology (d) Meteorology
2. Sunspots are
   (a) undulation on the solar surface
   (b) huge magnetic storms on the solar surface
   (c) places with adequate sunshine
   (d) places of tourist importance
3. The last great ice age existed in
   (a) Pliocene period (b) Pleistocene period
   (c) Miocene period (d) Triassic Period
4. The origin of the simplest life is attributed to
   (a) Proterzoic Era (b) Cambrian Era
   (c) Archaezoic Era (d) Carboniferous Period
5. How much time does a polar satellite require for an orbit of the earth
   (a) 2 hrs (b) 50 mints (c) 100 mints (d) 15 mints
6. Scientific name of Guinea grass
   (a) *Panicum miliaceae* (b) *Panicum maxi*
   (c) *Pennisetum purpureum* (d) *Pennisetum typhoides*
7. Sugar % in sorghum is
   (a) 10 (b) 17 (c) 6 (d) 11
8. Tirak is related to which crop
   (a) Maize (b) Chick pea (c) Cotton (d) Sunflower
9. Average loss of production due to weed in least developed countries is
   (a) 25% (b) 15% (c) 50% (d) 40%
10. The author of "Scientific Weed Management" is
    (a) Jethro Tull (b) V.S. Rao (c) O.P. Gupta (d) Crafts

11. The degree of uniformity of the occurrence of individuals of a species with in an area is called
    (a) Relative Dominance (b) Absolute Density
    (c) Relative frequency (d) Absolute frequency
12. Indian Society of Agronomy was established on
    (a) 1955 (b) 1960 (c) 1946 (d) 1974
13. The weed with maximum number of seeds/ plant is
    (a) *Ambrosia* (b) *Parthenium*
    (c) *Cyperus rotundus* (d) Lambs quarters
14. At field capacity, pF is
    (a) 2.54 (b) 4.2 (c) 1.0 (d) 10
15. Stomatal closing type of anti - transpirant is
    (a) Hexadecanol (b) phosphor (c) Adol - 52 (d) Kaolin
16. "Punjab" is a variety of
    (a) Rice (b) Wheat (c) Linseed (d) Pea
17. Dew Gauge was developed by
    (a) Dr. Duvdevani (b) Robertson (c) Pruit (d) Kush
18. Tilth index is varies from
    (a) 0-1 (b) 1-2 (c) 2-5 (d) less than zero
19. Poor's man root crop is
    (a) Potato (b) Tapioca (c) Beet (d) Raddish
20. Soil sickness is caused by
    (a) Maize (b) Jowar (c) Linseed (d) Both B & C
21. Gene responsible for high lysine & tryptophan content in maize
    (a) Opaque -2 (b) Flurry - 2 (c) Norin - 10 (d) Both A & B
22. Origin of niger is
    (a) Ethiopia (b) Mexico (c) Asia Minor (d) Abbysinia
23. Super rice variety is
    (a) Lunisree (b) Jaya (c) IR-8 (d) Ratna
24. Kharif, Rabi and Zaid words belong to
    (a) Urdu (b) Latin (c) Arabic (d) Greek
25. Natural Farming was advocated by
    (a) Gilbert (b) Fakuoka
    (c) Lawes (d) Norman E. Borlaug

26. Weed is used for indicator of NO_2

(a) *Stellaria media* (b) *Brassica kaber*

(c) *Cyperus rotundus* (d) *Argemone mexicana*

27. Cultivation of ragi begun in __________ BC. ()

(a) 2200 (b) 1800 (c) 7500 (d) 1200

28. Tift 23 D_2A is related to __________ crop.

(a) Wheat (b) Rice (c) Maize (d) Bajra

29. One cumec of water is equivalent to __________ cubic feet / sec of water.

(a) 33.21 (b) 40.31 (c) 35.31 (d) 25.31

30. Rice flower has __________ number of stamens.

(a) 4 (b) 6 (c) 3 (d) 9

31. Nurient mobility concepts was coined by-

(a) Hesse (b) Alexander Fleming

(c) Bray (d) Iwanowsky

32. Area under pasture cum grazing land –

(a) 11.30 mha (b) 14.63 mha (c) 3.6 mha (d) 75 mha

33. Storage substance for sunflower is

(a) Phaseolin (b) Inulin (c) Glutenin (d) Nodulin

34. ACRIP on sugar beet is at

(a) Srinagar (b) New Delhi (c) Simla (d) Pantnagar

35. Uniformity coefficient is introduced by.

(a) Bray (b) Christianen (c) Gilbert (d) Robertson

36. Crops having highest area under sprinkler irrigation system in India is

(a) Cereal (b) plantation crops

(c) Vegetables (d) Fruits

37. Boric acid contains ——-% B

(a) 26 (b) 21 (c) 18 (d) 24

38. Niger oil contains __________ % linolinic acid.

(a) 48 (b) 45-55 (c) 33-44 (d) 20 - 26

39. The best temperature for retting of sun hemp is

(a) 18 °C (b) 34 °C (c) 21-27 °C (d) 42 °C

40. White bud of maize is caused due to deficiency of

(a) Zn (b) Ca (c) Mo (d) Fe

41. Clark - 63 is a variety of

(a) Red gram (b) Soybean (c) Gram (d) Sunflower

42. The acid equivalent of urea is
(a) 129 (b) 110 (c) 80 (d) 93
43. Brix index in sugarcane indicates -
(a) Any other (b) Glucose % (c) Sucrose % (d) TSS
44. The % C content in organic matter is
(a) 58 (b) 62 (c) 33 (d) 50
45. Deficiency symptoms of N is
(a) Uniform yellowing of leaves including veins without dead spots on older leaves
(b) Uniform yellowing of leaves excluding veins without dead spots on older leaves
(c) Uniform yellowing of leaves including veins with dead spots on older leaves
(d) Uniform yellowing of leaves excluding veins with dead spots on older leaves
46. Poultry manure is richer in
(a) P (b) K (c) N (d) Ca
47. Pressmud cake is applied to reclaim
(a) Alkali soil (b) Acid soil (c) Saline soil (d) None of these
48. Leaching of nitrates cause
(a) air pollution (b) soil pollution (c) Noise pollution (d) water pollution
49. Rice with pericarp is called
(a) white riec (b) brown rice (c) pink rice (d) grey rice
50. Plank law is related to
(a) Diffusion (b) Radiation
(c) Saturated water flow (d) Wind velocity
51. Available Mo in soils is extracted by
(a) Citric acid (b) Ammonium acetate
(c) Hot water (d) Ca - chloride
52. Chromosome no of Sesamum indicum (2n)
(a) 26 (b) 32 (c) 64 (d) 42
53. Maximum productivity of sesame in India is in
(a) Orissa (b) T.N. (c) Karnataka (d) W.B.
54. Maximum productivity of mustard in India is in
(a) Punjab (b) Haryana (c) Rajasthan (d) W.B.
55. The inflorescence in castor is
(a) Raceme (b) panicle (c) spikelets (d) ear
56. Optimum temp. (°C) for growth & development of castor ranges between
(a) 31-37 (b) 25-30 (c) 20-26 (d) 16-20

57. Among the oilseed crops, castor ranks in India is
    (a) 8th (b) 6th (c) 9th (d) 5th
58. Symptoms of tip burn in plants can be induced by
    (a) Pathogens (b) Air Pollution (c) Salinity (d) Ca deficiency
59. Black heart of celery is due to deficiency of
    (a) Ca (b) Mg (c) Mo (d) Cu
60. Of the micronutrients, which one is essential for translocation of sugars in plant
    (a) B (b) Mo (c) Zn (d) Mn
61. Optimum age of seedlings for transplanting of *N. rustica* tobacco is —— weeks
    (a) 6-7 (b) 7-9 (c) 5-6 (d) 4-5
62. Heavy N application ——— of chewing, bidi type of tobacco
    (a) Increases yield and decreases quality
    (b) Increases quality and decreases yield
    (c) Increases both yield and quality
    (d) Decreases both yield and quality
63. The fruit of tobacco is called
    (a) Capsule (b) pod (c) siliqua (d) caryopsis
64. Sand drown in tobacco is due to
    (a) Cl (b) Ca (c) Mg (d) B
65. During drought conditions the deficiency occurs in cotton crops is
    (a) Cl (b) Ca (c) Mg (d) B
66. The area of major irrigation project is
    (a) >5000 ha (b) >15000 ha (c) >10000 ha (d) >20000 ha
67. Plant growth regulator used for ripening of seedless grape is
    (a) Glyphosate (b) NAA (c) GA (d) GA3
68. Plant growth regulator used for sucker control in tobacco is
    (a) Glyphosate (b) Coconut oil
    (c) MH (d) None of the above
69. Terai soils are deficient in
    (a) Zn (b) N (c) P (d) K
70. The crop which is grown through out the year in U.P. is
    (a) Maize (b) Black gram (c) Rice (d) Groundnut
71. The nature farming developed in
    (a) France (b) Japan (c) USA (d) Brazil

72. Transgenic crop having maximum area in the world
    (a) Maize (b) cotton (c) Soybean (d) Rice
73. Wilting of tobacco due to water logging is known as
    (a) Flopping (b) Scalding (c) Sponging (d) None of these
74. The ratio of water deficit and PET is known as
    (a) MDI (b) Aridity index (c) Stress index (d) PET index
75. Panicle initiation of wheat starts at
    (a) 21-25 DAS (b) 45 DAS (c) 75 DAS (d) 40 DAS
76. Fruit of wheat is known as
    (a) Ear (b) Panicle (c) Caryopsis (d) Grain
77. Test wt of *Phalaris minor* is
    (a) 2g (b) 4g (c) 6g (d) 10g
78. State known as "Bowl of Rice"
    (a) Punjab (b) UP (c) WB (d) Chhattisgarh
79. 1st hybrid variety of aromatic rice in world is
    (a) CORH 1 (b) PRH 1 (c) Pusa RH 10 (d) PHB71
80. 2,4-D resistant cotton gene is
    (a) Tift 23A (b) BIL 3B (c) tfd -A (d) None
81. The Phenolic compound present in cotton seed is
    (a) Gossypol (b) HCN (c) Glycosine (d) Glucosinolate
82. The N% in groundnut cake is
    (a) 5% (b) 7-8% (c) 2.5% (d) 1.5%
83. Glomerule is associated with
    (a) Sugarcane (b) Tobacco (c) Sugarbeet (d) Potato
84. Protein content in gram is
    (a) 21% (b) 23.5% (c) 25% (d) 24%
85. Black heart of potato is due to
    (a) O2 deficiency (b) Alkaline soil (c) Pathogen (d) Water stress
86. Glycoside found in Lucern is
    (a) Solanin (b) Leusine (c) Saponin (d) Isoleusne
87. Seed rate of lucern is ——— kg/ ha
    (a) 10-15 (b) 6-8 (c) 40-50 (d) 20-25
88. The chemical found in barseem is
    (a) Niacin (b) Saponin (c) Leusine (d) Iso-leusine
89. Plant P can be estimated on UV-VIS spectrophotometer at wavelength of

(a) 420 nm (b) 540 nm (c) 640 nm (d) 660 nm

90. As per WHO, the safe limit of nitrate content in ground water is —— mg/l
(a) 10 (b) 15 (c) 3 (d) 5

91. CEC of Illite is
(a) 100-200 (b) 50-60 (c) 15-40 (d) >200

92. Sunhemp is most widely used in India for
(a) Fibre (b) Fodder
(c) Green manuring (d) Minor vegetable

93. The term "Horticulture" is —— word
(a) Greek (b) Latin (c) Arabic (d) French

94. Blossom end rot of tomato is due to deficiency of
(a) Ca (b) B (c) Mo (d) Cu

95. The major importer of Mushroom from India is
(a) Canada (b) Brazil (c) UAE (d) USA

96. The Union Cabinet gave its approval for the establishment of *Food Legumes Research Platform* (FLRP) at.
(a) Rajasthan (b) U.P
(c) Madhya Pradesh (d) W.B.

97. International Centre for Foot and Mouth Disease (ICFMD) was established at
(a) Karnal, Haryana (b) Arugul, Bhubaneswar
(c) Sehore, Madhya Pradesh (d) None of these

98. MS Swaminathan pitched for the preparation and operation of which code of action for flood hit regions?
(a) Flood Code (b) Operation Flood
(c) Flood Scheme (d) None of the above

99. Field trials of GM Mustard are complete. If approved, it will become the ____ genetically modified food crop to be grown commercially in the country.
(a) First (b) Second (c) Third (d) Fourth

100. which microorganism is not involved in Sulphur cycle –, desulfuricans
(a) Thiobacillus (b) Arthrobacter (c) Desulfovibrio (d) Bacillus

Match the following:

	Weed		Wt of 1000 seeds (g)
I	Barnyard grass	a	22
II	Yellow nutsedge	b	0.70
III	Rice	c	1.4
IV	Bathua	d	0.19
V	Kochia	e	0.85

1. I - c, II - d, III - a, IV - e, V - b
   I - c, II - d, III - a, IV - b, V - e
   I - d, II - c, III - a, IV - b, V - e
   I - c, II - d, III - b, IV - a, V - e

	Crop		Criteria for harvesting of crops
I	Rice	a	28 - 35 days after flowering
II	Sorghum	b	35 - 40 days after flowering
III	Pearl millet	c	40 days after flowering
IV	Red gram	d	25 - 30 days after tasseling
V	Maize	e	32 days after flowering

2. I - e, II - c, III - b, IV - a, V - d
   I - c, II - e, III - a, IV - b, V - d
   I - e, II - c, III - a, IV - b, V - d
   I - e, II - d, III - b, IV - a, V - c

	Common name of weeds		Scientific name of weeds
I	Smooth joyweed	a	*Fimbristylis littoralis*
II	Cattail	b	*Biophytum sensitivum*
III	Broad Leaf Fig	c	*Centella asiatica*
IV	Life plant	d	*Alternanthera paronychioides*
V	Asiatic pennywort	e	Penisetum pedicillatum

3. I - d, II - e, III - a, IV - b, V - c
   I - b, II - e, III - a, IV - d, V - c
   I - d, II - e, III - c, IV - b, V - a
   I - d, II - b, III - a, IV - e, V - c

	Common name of weeds		Scientific name of weeds
I	Climbing dayflower	a	*Ipomoea cairica*
II	False amaranth	b	*Amaranthus spinosus*
III	Cairo morning glory	c	*Vicia sativa*
IV	Spiny pigweed	d	*Commelina diffusa*
V	Common vetch	e	*Digera arvensis*

4. I - d, II - b, III - a, IV - e, V - c
   I - d, II - e, III - a, IV - b, V - c
   I - b, II - e, III - a, IV - d, V - c
   I - d, II - b, III - a, IV - c, V - e

	Crop		Origin
I	Oat	a	China
II	Tobacco	b	South Africa
III	Soybean	c	Brazil
IV	Bajra	d	Asia Minor
V	Groundnut	e	South America

5. I - d, II - a, III - b, IV - e, V - c
   I - d, II - e, III - c, IV - a, V - b
   I - e, II - d, III - a, IV - b, V - c
   I - d, II - e, III - a, IV - b, V - c

	Crop		Water requirement
I	rice	a	400
II	wheat	b	153
III	mustard	c	1200
IV	pulse	d	600
V	Groundnut	e	300

6. I - a, II - a, III - c, IV - e, V - d
   I - c, II - d, III - b, IV - a, V - e
   I - c, II - d, III - b, IV - e, V - a
   I - c, II - a, III - b, IV - e, V - d

	weeds		**special organs**
I	Baloon	a	*Anemone*
II	wings	b	*Calotropis*
III	pappus	c	*Acer macrophylla*
IV	comose	d	*Physalis minima*
V	Feathery	e	*Tridex*

7. I - d, II - c, III - a, IV - e, V - b
   I - d, II - c, III - e, IV - a, V - b
   I - c, II - b, III - e, IV - a, V - d
   I - d, II - c, III - e, IV - b, V - a

	Fertilizers		**S Nutrient %**
I	K- sulphate	a	11.4
II	Ferrous sulphate	b	0.6
III	Urea gypsum	c	17.5
IV	Epsom salt	d	18.8
V	Cu - sulphate	e	13

8. I - d, II - c, III - b, IV - e, V - d
   I - c, II - d, III - e, IV - b, V - d
   I - c, II - d, III - b, IV - e, V - d
   I - c, II - d, III - b, IV - d, V - e

	Fertilizers		**Mn Nutrient %**
I	Mn - sulphate	a	63
II	Mn - oxide	b	12
III	Manganous oxide	c	10
IV	Mn - chelates	d	26
V	Mn - frits	e	41

9. I - a, II - d III - e, IV - b, V - c
   I - d, II - a, III - e, IV - b, V - c
   I - d, II - a, III - b, IV - e, V - c
   I - d, II - c, III - e, IV - b, V - a

	Crops		pH
I	rice	a	5-7.5
II	barley	b	4-6
III	cotton	c	5-5.5
IV	potato	d	6-7.5
V	oat	e	5-6.5

10. I - b, II - d, III - e, IV - c, V - a
    I - d, II - a, III - e, IV - b, V - c
    I - d, II - a, III - b, IV - e, V - c
    I - a II - d, III - e, IV - c, V - b

	Nutrients		Sources
I	B	a	Augite
II	Mo	b	Tourmaline
III	Cl	c	Olivine
IV	Cu	d	Apatite
V	Zn	e	Chalcopyrite

11. I - b, II - d, III - e, IV - c, V - a
    I - d, II - a, III - e, IV - b, V - c
    I - d, II - a, III - b, IV - e, V - c
    I - b, II - c, III - d, IV - e, V - a

	Crop		N fixed (kg/ha)
I	Groundnut	a	9-125
II	cowpea	b	49-450
III	soybean	c	50-66
IV	greengram	d	119-140
V	Black gram	e	27-206

12. I - e, II - a, III - b, IV - c, V - d
    I - d, II - a, III - e, IV - b, V - c
    I - d, II - a, III - b, IV - e, V - c
    I - b, II - c, III - d, IV - e, V - a

	Nutrient deficiency		Indicator plant
I	S	a	Cotton
II	Zn	b	Citrus
III	Mo	c	Tea
IV	N	d	Cauliflower
V	K	e	Maize

13. I - e, II - a, III - b, IV - c, V - d
    I - d, II - a, III - e, IV - b, V - c
    I - c, II - b, III - d, IV -e , V - a
    I - b, II - c, III - d, IV - e, V - a

	Herbicide		Trade name
I	Isoproturon	a	Sencor
II	Nitrofen	b	Dawpan
III	Propanil	c	Tok -E -25
IV	Metribuzin	d	Arelon
V	Dalapan	e	Stomp F 34

14. I - e, II - a, III - b, IV - c, V - d
    I - d, II - c, III -e , IV - a, V - b
    I - c, II - b, III - d, IV -e , V - a
    I - b, II - c, III - d, IV - e, V - a

	Plant nutrients		Close association
I	K	a	older leaves
II	Ca	b	young leaves
III	B	c	Heart rot (sugarbeet)
IV	Mo	d	Whiptail (cauliflower)
V	water	e	Jonathan spot (apple)

15. I -a , II - b, III - c, IV -d , V - e
    I - d, II - c, III -e , IV - a, V - b
    I - c, II - b, III - d, IV -e , V - a
    I - b, II - c, III - d, IV - e, V - a

Fill in the blanks:

1. If 90m area is covered while spraying a herbicide solution of 3.6 litres with a swath width of 1m, the volume of spray is ___________ litre.

2. The value of correlation coefficient varies from between ____________.
3. The earth crust contains on an average ____________ % P.
4. In most of the soil, the heat capacity ranges between ____________ cm^3.
5. The conversion of fertile soil in desert by wind erosion is known as ____________.
6. The family of sugar beet is ____________.
7. Neutralizing value of burnt lime is ____________.
8. Nearly ____________ % of water absorbed by the plants is lost in transpiration.
9. National Research centre for Agro-forestry at Jhanshi was established in the year ____________.
10. The required Fe concentration for normal plant growth is ____________ ppm.
11. The K_2O content in blood meal is ____________ %.
12. Redox potential of highly reduced soil is ____________ mv.
13. For nursery bed preparation of rice, the water requirement is ____________ mm.
14. The first miracle rice variety in India is ____________.
15. Amylose is determined by ____________ test.

References

1. A. Zaman and Md. Hedayetullah 2018 Agricultural Sustainability for crop Production, pp 247-263, In: A Zaman and S Maitra (Editors), Cutting Edge Technology for Agricultural Sustainability: New India Publishing Agency, New Delhi, ISBN: 978-93-87973-28-2.
2. A. Zaman and Md Hedayetullah 2019 Farming System and Sustainable Agriculture, Agrotech Publishing Academy, ISBN-978-81-8321-508-4.
3. Agrarian Transformation. Interaction between Traditional and Modernity, p. 38.
4. Agricultural geography an overview – Shodhganga shodhganga.inflibnet.ac.in/bitstream /10603 /107229/8/08_chapter%201.pdf.
5. Avinash S, Monoj S, Sheelawati M, and Nirupa K 2018 Present and Past Status of Indian Agriculture, *Agri Res & Tech: Open Access Journal,* **17(4)** 556040, DOI: 10.19080/ARTOAJ.2018.17.556040.
6. Deshpande T 2017 State of Agriculture in Indian, PRS.
7. DES, DAC& FW (Directorate of Economics & Statistics, Department of Agriculture, Cooperation & Farmers Welfare's), 2018 "Agricultural Situation in India". http//eands.dacnet.nic.in/publication.htm.
8. Ghosh PK, and Sahoo Bijoylaxmi 2011 Indigenous Traditional Knowledge, Orissa Review January -2011, Members of Orissa Secretariat Data Entry Operators' Association greeting with Shri Bijoy Kumar Patnaik, Chief Secretary on New Year's Day at Secretariat on 1.1.2011.
9. Hand book of Agriculture 2007 published by ICAR.
10. Hussain Majid 1982 Crop combination in India a Study, Concept Publishing Company, New Delhi.
11. http://eands.dacnet.nic.in
12. https://www.hort.purdue. edu/ newcrop/ Hort_306/text/lec05. pdf.
13. Ibid., p. 325 & 561.
14. Pandey MM 2009 Indian Agriculture – An Introduction, Submitted to Fourth Session of the Technical Committee of APCAEM 10-12 February, 2009, Chiang Rai, Thailand, *Central Institute of Agricultural Engineering* Bhopal, India.
15. Rudradatta KPM and Sundaram 2013 *Indian Economy*, p - 560.

16. Kareemulla K, and Ravichandran S 2018 105th FoCARS, ICAR-NAARM, Rajendranagar, Hyderabad, Telangana, India.
17. Manorama Year Book 2001 Growing the Food We Need.
18. Md. Hedayetullah, and Parveen Zaman 2018 Forages of the World, Major Forage Crops. CRC press, Taylor and Francis Group, Inc., New Jersey, USA: ISBN: 13:978-1-77188-684-0.
19. Rudradatta KPM and Sundaram New Agricultural Strategy and Green Revolution, p - 560.
20. Prakash N, Roy SS, and Ngachan SV 2013 Role of ITK in Conservation Agriculture: Blending Indigenous and Scientific Knowledge, ICAR Research Complex for NEH Region, Umiam – 793103, Meghalaya
21. Randhawa MS A History of Agriculture in India, **1** p - 23.
22. Jagadish S India, A Comprehensive Systematic Geography, p - 32.
23. State of Indian Farmers: A Report the Centre for the Study of Developing Societies (CSDS), Delhi conducted between December 2013 and January 2014.
24. Janardan S 2015 Teaching Manual on Introductory Agriculture (Ancient heritage, Agricultural scenario Gender equity in Agriculture), Department of Agronomy, Forages and Grassland Management, College of Agriculture, CSK Himachal Pradesh Krishi Vishvavidyalaya, Palampur 176062 India : College of Agriculture CSK HPKV, Palampur.
25. The World Book Encyclopedia. A. Vol. 1, p. 145. 147.
26. The New Encyclopedia. Vol. 1, p. 325. & 337-339.
27. Vijayalakshmi K and Sridhar S 1999 Pesticide Monitor, CIKS Chennai, **8** No. 3.
28. www. agricoop.nic.in
29. www.agrimoon.com/wp-content/ uploads/ Introductory- Agriculture.pdf.
30. www.agrimoon.com/introductory-agriculture-icar-ecourse-pdf-books/
31. www.icar.org.in
32. www. jau.in /coa / download/UG_Syllabus.pdf.
33. Yojana August 15 1993 Article – Agrarian Transformation, p - 40.
34. Zaman A and Maitra S 2019 Agricultural Heritage, Palmview publishing LLP, 189E, Kasba Road, Kolkata-700049, WB, India

Related Books

Titles	Authors
— A Grain Saved is a Grain Produced	S. Mohan
— Aadarsh Van Podhshala Sthapana evam Prabandhan (Hindi)	AFRI
— Agricultural Extension A Glimpse	S. Jananai
— Agricultural Microbiology	R.P. Pareek
— Agricultural Salinity Assessment and Management	K.K. Tanji
— Agriculture and Fertilizers	O.C. Bockman
— Agriculture Teachers Manual	NFFA
— Agro Enterprises for Empowering Farm Women	T. Rathakrishnan
— Agro-Entrepreneurship	S.R. Panigrahy
— Agrochemicals in Plant Disease Management	N.G. Ravichandran
— Aushdhiya Podhya Krishi Avam Upyog (Hindi)**	J. P. Tiwari
— Bhu Suchna Sateek Krishi Evam Nano Taknik (Hindi)	Ajay Srivastava
— Climate Agriculture and Man Shrinking Biodiversity and Sustainability	V.K. Sehgal
— Commercial Beekeeping	Dharm Singh
— Communication Skills and Personality Development	J.R. Kadam
— Competitive Nematology	N.G. Ravichandra
— Comprehensive Laboratory Manual of Life Science	Mamta Baunthiyal
— Crop Management *2nd Ed*	U.S. Walia
— Crop Production in Salt Affected Soils	I.C. Gupta
— Detection and Diagnosis of Plant Diseases	A. K.Meena
— Disaster Perception and Management	T.S. Chouhan
— Disease Problems in Vegetable Production *2nd Ed*	Satish K. Gupta
— Diseases of Vegetable Ornamental and Spice Crops	Satish K. Gupta
— Drainage Engineering: Principles and Practices	S.K. Gupta
— Drought Proofing and Water Resource Management	T.S. Chouhan
— Dryland Resources and Technology (Vol. 8)	L.L. Somani
— Dryland Technology *2nd Ed*	M.L. Jat
— Entrepreneurship Development and Communication Skills	R.R. Chole
— Entrepreneurship in Farming**	D. Khan
— Farm Mechanization for Production	D. Khare
— Farming System and Sustainable Agriculture	S.S. Walia
— Faslo Ke Rog: Parichay Evam Praband (Hindi)	Ritu Mawar
— Food Crops: Evolution, Diversity and Advances	D.A. Patil
— Forest Tree Seeds: Handbook	S.S. Sagwal
— Fundamentals of Agriculture *2nd Ed.* (Vol. 1-2) (Set)	R.L. Arya
— Fundamentals of Agronomy	R.L. Arya
— Fundamentals of Human Ethics and Agriculture Extension	D.K. Meena
— Fundamentals of Organic Farming and Gardening: An Instructors Guide	A. McGregor
— Fundamentals of Soil Science	R.L. Arya
— Fundamentals of Soil Science	A. Rathinasamy

Related Books

Titles	***Authors***
— Fundamentals of Vegetable Crop Production	Beena Nair
— Indian Agri & Agribus Management *3rd Ed*	Smita Diwase
— Indiras Objective Agricultural Biotechnology *2nd Ed*	R.L. Arya
— Indiras Objective Agricultural Extension	R.L. Arya
— Indiras Objective Agronomy *3rd Ed*	R.L. Arya
— Insect Pathology Text Book and Practical Manual	N.K. Maru
— Integrated Plant Disease Management	R.C. Sharma
— Introduction to Crops of India *2nd Ed*	N.R. Das
— Irrigation Resources	S.B. Verma
— Key to Success in Agriculture	B.N. Kalsariya
— Krishi Jagat *2nd Edition*: JET Evam Krishi Paryavekshak Pariksha Upyogi (Hindi)	M.L. Jat
— Management of Diseases and Insect-Pests of Important Vegetables **	S.K. Gupta
— Manual on Fundamentals of Agronomy	L.K. Jain
— Marda Vigyan ky Mulbhut Sidhant (Hindi) -	H. P. Parewa
— Masala Faslo Ki Vaigyanik Kheti (Hindi)	Kamini Kumari
— Nanofertilizers: Challenges and Prospects**	J.C. Tarafdar
— Nanotechnology	J.C. Tarafdar
— New Vistas of Organic Farming *2nd Ed*	Mukund Joshi
— Objective Agribusiness Mgmt. *2nd ed*	S.R. Panigrahy
— Objective Seed Science and Technology *2nd Ed.***	K. Vanangamudi
— Organic Farming Theory & Practice	S.P. Palaniappan
— Practical Manual on Basic Agronomy *2nd Revised Ed*	N.R. Das
— Principles of Organic Farming: Textbook**	P. L. Maliwal
— Quinoa *(Chenopodium Quinoa Willd)*	Dharm Singh
— Remote Sensing and GIS Applications: A Starter Guide	D.C. Joshi
— Samanvit Krishi Pranali Prabandan (Hindi)	L.R. Meena
— Sasya Vigyan Ka Moolbhut Gyan (Hindi)	Ram Pyare
— Sasya Vigyan Ke Sidhant (Principles of Agronomy) (Hindi)	R.L. Arya
— Seed Technology *2nd Revised & Enlarged Edition*	D. Khare
— Seed Technology in The Tropics	D.B. Mackay
— Soil Guide: A Handbook for Understanding and Managing Agricultural Soils	Geoff Moore
— Spice Crops Production Technology	P.C. Das
— Sustainable Agriculture	R.S. Meena
— Sustainable Development of Dryland Agriculture in India	R.P. Singh
— Sustainable Dryland Farming	A.M. Dhopte
— Textbook of Rainfed Agriculture and Watershed Management	P.L. Maliwal
— Vegetable Crops at a Glance	S.K. Tyagi
— Weeds Identifications and Medicinal Use *2nd Ed***	U.S. Walia